STEM Careers

An indispensable guide to opportunities in science, technology, engineering and maths

Liz Painter

STEM Careers: An indispensable guide to opportunities in science, technology, engineering and maths

This 2nd edition published in 2024 by Trotman, an imprint of Trotman Indigo Publishing Ltd, 18e Charles Street, Bath BA1 1HX

© Trotman Indigo Publishing Ltd 2024

Author: Liz Painter

1st edition: Paul Greer

British Library Cataloguing in Publication Data
A catalogue record for this book is available from the British Library.

ISBN 978 1 911724 18 6

All rights reserved. This book is sold subject to the condition that it shall not, by way of trade or otherwise, be lent, resold, hired out or otherwise circulated without the publisher's prior written consent in any form of binding or cover other than that in which it is published and without a similar condition including this condition being imposed on the subsequent purchaser. No part of this publication may be reproduced, stored in a retrieval system or transmitted in any form or by any means, electronic and mechanical, photocopying, recording or otherwise without prior permission of Trotman Indigo Publishing.

Every effort has been made to trace copyright holders and to obtain their permission for the use of copyright material. The publisher apologises for any errors or omissions, and would be grateful to be notified of any corrections that should be incorporated in future editions of this book.

Cover design by Frasco Design

The authorised representative in the EEA is Easy Access System Europe Oü (EAS), Mustamäe tee 50, 10621 Tallinn, Estonia.

Printed and bound in the UK by 4Edge Ltd, Hockley, Essex

 All details in this book were correct at the time of going to press. To keep up-to-date with all the latest news and updates and to access the online resources that accompany this book, use the QR code or visit trotman.co.uk/pages/stem-careers-resources

Contents

How to use this book	vii
About the author	viii
Acknowledgements	ix

1| What does STEM careers mean? — 1
Introduction — 1
What does 'career' mean? — 1
What does STEM mean? — 1
The fourth industrial revolution — 4
More information about . . . Artificial intelligence — 5
STEM is for all — 5
Global and local issues — 6
Collaboration is key — 7
Hero in STEM: Professor Hannah Fry — 7
Conclusion — 8
Reflection about Chapter 1 — 9
Career story: Lauren, Kiln Engineer — 10

2| Is STEM for me? — 13
Introduction — 13
Are there people like me in STEM? — 13
Hero in STEM: Dame Dr Maggie Aderin-Pocock — 15
Jobs at all levels — 16
STEAM — 16
Charity organisations — 17
Non-STEM roles in STEM organisations — 18
STEM skills required in all jobs — 18
Green: Skills, jobs, careers and economy — 19
Conclusion — 19
Reflection about Chapter 2 — 20
Career story: Molly, Multi-Platform Podcast Producer — 21

3| STEM sectors (making things) — 23
Introduction — 23
Grouping the STEM sectors (making things and helping others) — 23
More information about . . . UK Armed Forces careers — 24
Six STEM sectors involved in making things — 25
More information about . . . Small to medium enterprises (SMEs) — 29
Hero in STEM: Professor Dame Sarah Gilbert — 32
Creative industries — 32
Conclusion — 34
Reflection about Chapter 3 — 34
Career story: Ethan, Engineering Degree Apprentice — 35
Career story: Josh, Mechanical Engineer Apprentice — 37

Contents

4|	**STEM sectors (helping others)**	**39**
	Introduction	39
	Six STEM sectors involved in helping others	39
	Hero in STEM: Hamza Yassin	42
	More information about . . . Sustainability, sustainable careers, careers in sustainability	42
	More information about . . . NHS graduate training opportunities	45
	The value different industries contribute to the UK economy	48
	Conclusion	50
	Reflection about Chapters 3 and 4	50
	Career story: Catriona, Systems Analyst (retired)	51
	Be your own careers researcher	**53**
5|	**Useful skills and personal qualities for STEM roles**	**55**
	Introduction	55
	What does the term 'skills' mean?	55
	Skills linked to knowledge (and reflection activities)	56
	More information about . . . Digital literacy	60
	Reflection on technical and digital skills	61
	Skills linked to self-awareness	61
	Hero in STEM: Sir James Dyson	63
	Reflection on self-awareness	64
	Skills linked to working with others (and reflection activities)	65
	Bringing skills together	68
	Values	69
	Reflection on work values	70
	Conclusion	71
	Collective reflections about Chapter 5	71
	Career story: Talal, Business Intelligence Analyst	72
6|	**STEM qualifications (making choices)**	**75**
	Introduction	75
	Making good career decisions	75
	Career steps	76
	School leaving age	76
	Study, work or both?	77
	Taking a gap year	77
	Unplanned events	78
	Changing pathways	79
	Comparing qualifications	79
	Conclusion	82
	Reflection about Chapter 6	83
	Career story: Ruaridh, Environment Forester	84
7|	**Study (with work experience)**	**87**
	Introduction	87
	What are transition stages?	87
	Why stay in education?	88
	Types of qualifications	88
	More information about ... Higher Technical Qualifications	92
	Choosing a university	95

	Funding further study	96
	Having a part-time job	96
	Conclusion	97
	Reflection about Chapter 7	98
	Career story: Hannah, Primary Care Pharmacist	99
8|	**Work (with study)**	**101**
	Introduction	101
	Why go straight into employment?	101
	Traineeships (England)	102
	National Vocational Qualifications (NVQs)	102
	Scottish Vocational Qualifications (SVQs) are work-based qualifications	103
	Apprenticeships	103
	Degree apprenticeships	104
	How to find an apprenticeship or traineeship	105
	Focus on an organisation: Specsavers	105
	Benefits of having a vocational qualification	106
	Is training the easy option?	106
	Graduate schemes	107
	Focus on an organisation: Direct Line Group	107
	Conclusion	108
	Reflection about Chapter 8	109
	Career story: Bonnie, Hearing Aid Dispenser Trainee	110
9|	**What do STEM job adverts tell us?**	**113**
	Introduction	113
	Types of jobs	113
	Dissecting a job advert	115
	Hero in STEM: Professor Sir John Holman	118
	Where can you find STEM job adverts?	118
	But I'm not ready for a job yet!	119
	On-site, remote or hybrid?	119
	Conclusion	120
	Reflection about Chapter 9	121
	Career story: Doyinsola, Area Network Controller	122
	Career story: Alex, Fast Particle Modeller	124
10|	**Preparation and planning for a STEM career**	**127**
	Introduction	127
	Getting started	127
	Planning	128
	Personal statements and CVs	129
	Sources of STEM and career-related information	129
	People who can support you	130
	Helping yourself	131
	Work experience (or *experience of the workplace*)	132
	Is it all down to you? (Networking)	133
	Reviewing and adjusting plans	134
	Conclusion	134
	Reflection about Chapter 10	135
	Career story: Dionne, Process Engineer	136

Contents

11|	**For parents/carers**	**139**
	Introduction	139
	Young people managing their career	139
	Job fog	140
	Using this book	140
	Supporting your young person	141
	Home educated?	142
	Conclusion	142
12|	**For teachers**	**145**
	Introduction	145
	How you can use this book	145
	Other features of the book	145
	Other ways you can help your students	147
	Conclusion	148
	A–Z 60 additional STEM career ideas	149

How to use this book

You can read this book in the order it is written, or you can dip in and out of the sections you are interested in.

References are made to websites throughout the book along with the URL links.

There are special features in this book:

- Each chapter ends with a 'reflection' box that gives ideas to help you think about STEM careers and your own career decisions.
- There are interviews (we have called them 'Career stories') with people in their early STEM career. Please do read them, even if you think you are not interested in their job title. They come from across the UK and work in a range of sectors. All have good stories sharing about their education, making mistakes and getting into work and life in general!
- There are 'Fascinating facts' about current ideas and innovations in STEM throughout the book (e.g. insects for food and doodling leading to a major discovery). If something interests you, research it further to find out more information.
- Six 'Heroes in STEM' are included with a brief profile of what they do. You may not have heard of them all, but their work will probably affect yours and your family's day-to-day life. You can research the 'hero' further – where, and what, did they study? What similar courses are available at colleges or universities you are interested in?

About the author

Liz Painter is a freelance career development professional, has an MA in Careers Education and Coaching, and is registered with the Career Development Institute (CDI).

She was a science teacher for 24 years and led a range of STEM activities, including clubs, cross-curricular collaboration days and STEM in PSHCE programmes. Liz was invited to speak at two Inside Government conferences about encouraging girls into STEM careers.

After leaving teaching in 2019, she joined the Cheshire and Warrington LEP, where she supported schools with their careers education programme and facilitated employer engagement activities. In 2021, Liz began to concentrate on her freelance portfolio. She delivers training, helping teachers to bring career learning into the classroom. Liz has worked on projects for a number of organisations, including STEM Learning, developing career-related resources for educators and young people.

Acknowledgements

The author would like to thank the people who contributed their career stories to this book and the organisations and individuals who offered guidance about the content. In particular, the author would like to thank her family (especially Andy, Caitlin, Ewan, Sam and Jamie) for their support and suggestions for many of the chapters.

This book is dedicated to Mr G, a brilliant engineering teacher who helped many girls to see that STEM could be for them.

1 What does STEM careers mean?

Introduction

If you've picked up this book, you probably already know what the acronym STEM stands for, but just in case you don't, STEM represents science, technology, engineering and mathematics. You may not know what job you'd like in the future. Perhaps you already have an idea of what career you would like and you are hoping that this book will confirm your plans. Maybe you enjoy what you are learning in science, maths and technology lessons and want to find out more about possible courses you could study or jobs you could do in the future that are linked to your interests. Whatever your reasons for reading this book, you'll find suggestions to help you prepare for your next steps. You might even find some new career ideas to explore.

What does 'career' mean?

You may have heard people talk about careers, either their own career or encouraging you to think about your career plans. Throughout this book, career will be defined as 'an individual's pathway through life, learning and work' (Andrews and Hooley, 2018). That means you've already started to develop your career! The subjects you've chosen to study, and any part-time work and volunteering you do all contribute to your career development and understanding of the world of work. Your hobbies and interests may also feed into your career development, as they may provide inspiration for your future job ideas. As well as informing you about STEM careers, this book will help you to make good decisions about gaining relevant work experience, choosing what to study and applying for courses, apprenticeships and jobs.

What does STEM mean?

Science, technology, engineering and mathematics are involved in just about every part of our lives, for example, developing and producing the food we eat, the clothes we wear and producing the games, films and music that entertain us. STEM is at the heart of solving the local and global problems that we need to address. Developing new medicines and health care treatments and providing sustainable sources of energy and infrastructure such as roads, buildings and railways that can cope with future demands all involve STEM.

The four STEM subjects do not work in isolation from each other or other subjects! To give you an example, if you look up the *sound engineer job profile* you'll see that either music, physics or maths could be a stepping stone into this industry. *Archaeologist* is another good example, as computing, geography and history qualifications could help you on this career pathway.

However, as an introduction to the four areas of STEM, here is a quick look at each subject.

Science

At school this really means biology, chemistry and physics. Depending on which part of the UK you live in and how old you are, you may choose to focus on one or two of these sciences or you may study all three. Post-16, that is after GCSEs (England, Wales and Northern Ireland) and National 5s (Scotland), you have much more control over what you choose to study. You may choose A levels, or Highers and Advanced Highers in Scotland, in one or all of the three sciences. There are also vocational science qualifications you can study at college, such as laboratory science and health and social care. After school/college, there are very specific science degrees you can study at university such as biochemistry, astrophysics and pharmacology. Pathway choices will be covered in more detail in Chapters 6, 7 and 8. A person with a career in science will usually have focused on one area, but they will use their mathematical skills and probably a lot of technology. Read Hannah, the pharmacist's story in Chapter 7, for an example of a career in science.

For ideas about careers in science, visit Cogent's Science Careers Pathways website (www.sciencecareerpathways.com/home/).

Technology

The dictionary definition of 'technology' is the application of science for a practical purpose. Technology subjects at school may include information and communications technology (perhaps within computer science) and design and technology (Design and Manufacture, Fashion

1 What does STEM careers mean?

and Textiles in Scotland). In college, university and the workplace, technology can become very specialised. But even if you don't choose to study technology, you'll use it in your day-to-day life, from the phone you have beside you to the machines and devices you come across in the home, shops, hospitals and cars. In fact, just about every job involves using tech, so being *digitally literate* is a useful skill to have. Skills will be discussed in more detail in Chapter 5. A career in technology could involve creating and using tech, such as a web designer or software developer. Part of the work may involve designing and building things and this is engineering. Read Talal's career story in Chapter 5 about how he uses technology in his job as a business intelligence analyst.

Listen to the podcasts from Tech UK to learn more about the cutting edge of the UK's tech industry (www.techuk.org/what-we-deliver/podcast-page.html).

Engineering

At school and college most of the engineering courses you'll come across will probably be mechanical and electrical. You'll learn how science and maths can be used to make things, invent new things and solve problems. There are so many different types of engineer, and organisations have joined together to create the 'Meet the Future You' quiz to explore the different engineering careers. (Have a go at the quiz, www.mtfy.org.uk/)

Here are two different types of engineer. Civil engineers design and build roads, buildings and other infrastructure. Genetic engineers are usually based in laboratories and work with genes and DNA to improve health and food supply. Both will use technology to help them do their job and both are trying to make society better for people. Chapter 3 will look at different STEM sectors, including the main engineering sectors. Engineers like to take things apart to see how they work, as well as making things. So any practical experience is useful if you think a career in engineering may be for you, but be safe and ask for permission before you take the toaster apart! Read Dionne's story about being an engineer at Jaguar Land Rover, in Chapter 10.

Mathematics

I challenge anyone who says they can't do maths or thinks maths is boring! We all use maths throughout the day, from making sure we get a bargain at the shops to working out our schedule for the day. My hairdresser moans about not being able to do maths but can whip up a 20% solution of hair dye without even thinking about it! You're reading a book about STEM careers, so you're probably interested in science, technology or engineering and hopefully maths. When you're narrowing down your courses at post-16, try not to throw maths out too quickly. Maths underpins the other three STEM subjects, helping you to develop good numeracy, analytical and problem-solving skills. Maths matters – having these skills will be invaluable in whatever STEM career pathway you follow. Just take a look at a few job profiles and job adverts to see how often numerical skills come up. Read about Catriona's love of maths in her career story in Chapter 4.

For lots of examples of careers using maths, visit the Maths Careers website(www.mathscareers.org.uk/).

The fourth industrial revolution

A brief word about this term you may have heard. The 'fourth industrial revolution' relates to the rapid changes in technology and how this is affecting society. The scale and speed of this change are huge and this brings opportunities and risks.

Opportunities include using technological developments to solve local and global problems such as climate change. Advances in technology will help us to find new medicines and sustainable energy sources. Benefits also include the new training opportunities these developments bring. For example, car mechanics can 'up-skill' and learn to service and repair hybrid and electric vehicles.

However, the risks associated with new technology can affect people and society. Think about an elderly relative who feels they are being 'left behind' with phones and computers. People like you will help to solve these problems by developing inclusive ways for us all to access new technologies.

Some people are concerned that they may lose their job due to automation, for example, robots doing the work on a production line. However, new jobs will be created, such as building, maintaining and programming the robots! So, it is important to be willing to learn and develop throughout your career.

> **MORE INFORMATION ABOUT . . .**
>
> *Artificial intelligence (AI)*
>
> AI refers to the ability of a computer, or computer-controlled robot, to perform the tasks that a human can usually perform.
>
> AI technology involves the machine processing lots of data and learning. (A human would struggle to do this in the same amount of time.) From this, AI is able to analyse, make predictions and recommendations. (This is basically how chatbots such as ChatGPT work.)
>
> This is changing the world of work, and many routine tasks are now automated. Although this means that some low-skilled jobs may be lost, it also provides opportunities to learn new skills and how to work alongside AI.
>
> For more information, visit the Cloud Google website (cloud.google.com/learn/what-is-artificial-intelligence).

STEM is for all

People involved in STEM are going to be at the forefront of helping to make society a better place for everyone and solving local and global issues. To do this we need a range of people from across society to work in STEM-related sectors. Diversity, inclusivity and equality are more than buzzwords that employers and other organisations are using. Forget whatever stereotypes you think you know of, such as all chemists wear lab coats and engineering is a mucky job. We can't afford to lose any potential creativity and talent because people may think that a career in a certain STEM job isn't for them. A STEM workforce that represents society, in terms of gender, ethnicity, social and cultural background, sexuality, disability and neurodivergence, is important. There is a saying that 'you can't be what you can't see', and in some ways this is true. Hearing career stories from people who are like you can be influential in helping you to see different career pathways. Throughout this book, there will be STEM career stories from a range of people with different experiences. I hope you enjoy reading them. To start, read Lauren's career story at the end of this chapter to find out how she got from dog grooming to being a kiln engineer.

Global and local issues

The phrase 'global and local issues' has been used a few times, and these concepts are important to you. Your journey through education and into employment is intertwined with what is happening where you live and around the planet. Understanding the 'bigger picture' of the economy and politics and how they affect society may help you to make better career decisions. Think about who your big local employers are (where do 'everyone's' parents work?), what area, or sector, does this business work in? Is the business embracing the fourth industrial revolution or are people worried about their jobs? These are questions to consider before following what might be an easy, local pathway into employment.

In January 2020 the UK officially left the European Union. This has been called Brexit. The fallout from Brexit is still apparent as the UK's international trade has been affected. Less money is being invested in UK business, and the workforce has been altered with a shortage of labour in hospitality and retail. You may not think this directly affects you, but think back to big local employers who may be reducing the career opportunities in your area because they are choosing to invest elsewhere.

Since March 2020 the whole world has been affected by the Covid-19 pandemic, and the effects are ongoing. The personal and societal suffering was dreadful. The importance of healthcare and medical research, food supply, transport and logistics was at the forefront of everyone's minds. Political uncertainty in Europe during 2022 affected energy supplies in the UK. The government and many organisations are prioritising providing sustainable energy to meet the UK's needs. People with STEM jobs will help to find the solutions.

In fact, the changing local and global landscape will provide new and exciting career opportunities, both in the UK and other countries. A company may offer employees the chance to relocate to another country, for example, to work on a specific project. The pandemic forced companies to make home-working easier for staff. Developments in online platforms occurred rapidly, and this has changed the dynamics of working from home. Being aware of the changing world of work is important as in the future you may want the opportunity to work in another country or from home.

It isn't hard to keep abreast of local and global issues – watch the news, download a respected newspaper app and read the headlines. Listen to the stories friends, family and neighbours are telling about their work.

Collaboration is key

At the moment you may see your STEM subjects as separate. In the workplace you may have a very defined specialism, for example, designing new medicines you'll use a variety of subject knowledge (e.g. biology, chemistry, maths, IT) and work with people who have different specialisms (e.g. healthcare, business, legal). This range of subject knowledge and specialisms is required to design and test a new drug and safely deliver it to market. The drug designer is just one part of the bigger picture in this project. The skills and characteristics required to collaborate and work well with others are discussed further in Chapter 5.

If you think of anything that you use, many subject-specialists will have collaborated to achieve the finished product or service. From designing and manufacturing new foods to making the things in your home and the book you are reading, STEM knowledge, creativity and business awareness are required.

The world of work is interdisciplinary and understanding this can help you to have a successful STEM career. You may already have an idea of what STEM career you'd like or are still keeping your options open. Either way, it is worth spending time considering the subjects you could choose to study at the next stage in your education and how they may be of use to you in the future. Subject choices will be discussed in more detail in Chapters 6, 7 and 8.

 HERO IN STEM: PROFESSOR HANNAH FRY

In her own words from her website, Hannah Fry is a 'Mathematician, science presenter and all-round badass' (https://hannahfry.co.uk/).

Dr Fry is a professor in the Mathematics of Cities at University College London. Collaborating with geographers, physicists, computer scientists and architects, she studies patterns of human behaviour within towns and cities. This research is useful to a variety of organisations including urban planners, supermarkets and the police, for example, with transport and crowd behaviour. The interdisciplinary teams are working together, using data science and modelling to help create safer and better towns and cities.

Hannah is also a successful science communicator, writing best-selling books about mathematics and technology, and making radio and TV programmes, such as *Curious Cases of Rutherford and Fry* on Radio 4 and *The Secret Genius of Modern Life* on BBC 2.

> **Fascinating fact**
>
> Entomologists estimate that they have only discovered 1/10th of the world's insects. We all know that these bugs are important pollinators and decomposers, but did you know scientists and technologists are developing ways to grow and harvest them as a source of protein for humans?

Conclusion

This book is a guide to help you consider a range of STEM careers. You can read through the book in order or can dip into specific chapters when you require certain information. You can come back to the book as you progress through your education and make future decisions.

- **Chapter 2** will help you to understand if a STEM career is for you and also discuss why inclusivity and creativity are important.
- **Chapter 3** contains six STEM sectors that are involved with 'making things'. It includes information about 30 different types of jobs within these sectors.
- **Chapter 4** contains six more STEM sectors, which are involved with 'helping others'. Again, there are another 30 job titles for you to explore.
- **Chapter 5** explains 12 important skills valued by employers and how they can be developed. Work values are also discussed, and why thinking about these is important when considering future employers.
- **Chapter 6** focuses on making good career decisions and how academic and vocational pathways do not need to be separate.
- **Chapter 7** looks at different study routes and qualifications across all four home nations. The pros and cons of staying in education, and choosing universities and courses are also covered.
- **Chapter 8** focuses on heading into employment with training. Finding an apprenticeship and things to consider are also discussed. STEM graduate training schemes are also highlighted.
- **Chapter 9** walks you through a job advert, showing how this can be a useful exercise to find out more about STEM jobs. Understanding what is required to be successful in applying for a job can help you to plan and develop good career management skills.
- **Chapter 10** pulls together the ideas from each chapter, helping you to develop positive behaviours and plan for their transitions into a STEM career. Gaining work experience is also discussed.

1 What does STEM careers mean?

- **Chapter 11** is for parents, carers and other supportive friends and family, so they can help you to find out more about STEM careers and make good decisions as you progress from education into the workplace.
- **Chapter 12** will give teachers ideas and lesson resources to help them bring career learning into STEM lessons.

Of course, the person making the decision is you. Therefore, at the end of each chapter there are suggested reflective questions. These will encourage you to think about how you can use the information in that chapter, develop your own research to expand your knowledge about STEM careers and how to manage your career and plan your next steps.

The book is also designed to help you to make good decisions. If you start out in one direction and then change your mind, that's OK. But it should be because things have changed for you and not because you didn't have enough information to make a good decision.

People around you, such as family and friends, teachers and careers advisers, can help you gather information and make decisions. Meeting and speaking with employers is also a great way to find out about a job, the day-to-day work and career progression opportunities. You can also find out more about STEM careers from websites, and many are listed in the book.

 REFLECTION ABOUT CHAPTER 1

Think about why you are interested in science, technology, engineering and/or maths. What is it about these subjects and what you learn that made you pick up this book and read to the end of the first chapter? Keep a note of your thoughts as they'll be useful in Chapter 2.

Career story: Lauren

Kiln Engineer, Heidelberg Materials, North Wales

What is your job?

I am a kiln engineer, which is a type of production engineer. We use limestone, sand, pulverised fly ash and iron oxide to produce clinker, which is one of the ingredients for cement. First, we preheat the raw materials to $900^{\circ}C$, then they enter the even hotter kiln and the raw materials then form clinker.

I'm busiest when the plant is running, ensuring it is working as efficiently as possible. That means maximising clinker production, lowering fuel consumption while hitting key targets such as reducing carbon dioxide emissions.

When the plant is on a maintenance shutdown, I still have work to do. It is my job to supervise installing new refractory material that protects the steel inside the kiln, pre-heater tower and cooler. This is really important as temperatures above $400^{\circ}C$ can weaken the steel, and the kiln flame can reach temperatures up to $1,450^{\circ}C$!

Describe your career journey so far

I did not like high school, but I did enjoy learning science. I did triple science GCSE and A levels in biology, chemistry, physics and sociology. As I was good at science, I was pointed in the direction of medicine and studied medical science for a year. I didn't like this and decided it was not for me so I switched to a biology and conservation degree. I wanted an animal-based career and to be Steve Irwin, but as there were no crocodiles in Wales, I had to rethink my plans.

Since I was 16, I had part-time jobs in pet shops, and I stayed in animal care after I graduated and went into dog grooming, rising to management. But although I loved dogs, this job was not fun. As a manager, I had the difficult dogs to work with, and one day while moaning to my older brother when I was fed up with being bitten, he suggested I work for the company he was with, called Hanson. Hanson, now known as Heidelberg Materials, produces building materials.

I was 26 at the time. I didn't know anything about making asphalt (a type of road surfacing), but he encouraged me to apply for the asphalt operations job. I didn't get that job, but I must have made an impression at the interview because seven months later the company invited me back to interview for another job which I got.

They took a big risk as they had to train me to be an asphalt plant operator. I did this job for a year and loved it. I then became the asphalt plant supervisor and then I was made a district operations manager to build a mobile asphalt plant. I was then recruited to Hinkley Point C as a concrete production manager for Hanson, and I worked there for three years.

While at Hinkey Point C, I worked closely with a materials engineer and this made me want to train to be an engineer. A kiln engineer job came up with Hanson and I got it, but I negotiated the terms and persuaded the company to put me through an engineering degree alongside doing the role. They agreed as they had recruited me for the managerial skills I'd developed during my previous roles. This allowed me to develop my engineering skills and I finished my degree in mechanical engineering in September 2023.

What advice would give to someone still at school?

Don't be afraid to take risks and try something new. But be determined to succeed because there will be times, maybe three to six months in, when things will get hard. You'll think 'this is crazy, what I have done?' This is not the time to quit, dig deep, show up every day, as this time will pass. You'll build up the skill set and realise you do know what you're doing.

2 Is STEM for me?

Introduction

Think a career in STEM might not be for you? If you are interested in films, TV, theatre, dance, music, food, fashion, gaming or social media then think again. Perhaps you are passionate about supporting the underdog and social causes at local or global levels? Creativity and innovation are required to solve the problems on our doorstep and around the world. Creative minds are needed to make the music in the entertainment business, the marketing used in social media, sustainable food sources of the future and the responsible textiles in tomorrow's fashion. All involve some element of STEM.

Are there people like me in STEM?

Yes! Remember the World War II slogan 'Your country needs you'? It still applies today. The purpose of every job is to help that company achieve its goal. That could be making people better, designing more efficient machines or improving access to food, energy and water.

Employers understand that for the company to succeed, a broad range of skills, abilities, opinions and approaches to problem-solving are required. New ideas can be generated when a variety of people with different perspectives work together. Whatever your gender, ethnicity, academic ability, physical ability, social background, sexuality or neurodivergence, you can have a STEM career. Your skills, experiences, point of view and ideas are needed in STEM.

Gender

There has been a lot of work done in the last decade to encourage girls into STEM careers. There is also work to be done to challenge the stereotypes of typically 'female' careers, such as nursing and care work. There are no jobs that are gender-specific. Sometimes parents, teachers and friends may question your career ideas, but don't let that put you off. Use it as a way to discuss and explore your reasons for considering a particular career. Women in STEM has further information at www.womeninstem.co.uk/.

Ethnicity

The term 'ethnicity' refers to all groups except the white British group. Ethnic minorities include Asian or Asian British; Black British, Caribbean or African; multiple ethnic backgrounds; white minorities, such as Gypsy, Roma and Irish Traveller groups. People from all ethnic minority groups are underrepresented in STEM. There are many organisations whose mission is to provide support and mentors to help students from minority ethnic backgrounds explore careers in STEM. An example is Black British Professionals (https://bbstem.co.uk/).

Disability

The Equality Act (2010) makes it illegal for employers to discriminate against employees with a disability. Some disabilities can be 'hidden', such as sensory impairment and mental health conditions. If your employer is aware of your disability, they will be able to make reasonable adjustments to help you thrive at work. Employers value the unique perspective that people with a disability bring to the workplace. Often, they are able to see problems and offer solutions that others may overlook. The website for the global network of disabled scientists has further information at www.sciendis.org/.

Social background

Your social background is the type of family you come from and education you receive. Perhaps you'd like a career that no one else in your family or social circle has. This can be daunting as often those you seek advice from and trust may feel unable to guide and support you because they don't know anything about that job. This is when talking with a careers adviser can be really useful. Make an appointment to see them in school/college to discuss your career idea. They will give you impartial advice and help you to plan how to find out about the career pathways to get to your dream job. In 2 Science promote social mobility and diversity in STEM (https://in2scienceuk.org/).

LGBTQIA+

Many members of the LGBTQIA+ community want to be visible within the STEM community. When voices are heard and listened to, the workplace can become a safer and fairer space for all. There are many local and sector-specific LGBTQIA+ support groups for all areas of STEM. If there is a company you are interested in, have a look at their website to see how they promote inclusivity. For more information, visit https://prideinstem.org/.

Neurodivergence

People who are neurodivergent (e.g. are dyslexic, autistic or have ADHD) often have skills that employers value such as attention to detail, unique problem-solving skills and the ability to work autonomously, meaning they can work successfully by themselves. Many employers are embracing inclusivity and will support neurodivergent applicants during the recruitment process and in the workplace. Cacti is working towards neuroinclusive careers in STEM (www.cacti.org.uk/).

 HERO IN STEM: DAME DR MAGGIE ADERIN-POCOCK

Dame Maggie was born in London and always loved science and wanted to become an astronaut and go into space. Undiagnosed dyslexia, attending 13 different schools and being told a career in the caring profession was 'a more realistic aim' didn't put Maggie off pursuing her goal to work in the space industry.

Maggie completed a physics degree and PhD in mechanical engineering at Imperial College London. Her career so far has involved developing landmine detection equipment, satellites with instruments to monitor climate change and work on the James Webb Space Telescope technology.

Whatever their class, gender or background, Maggie's passion is to inspire and encourage young people to consider careers in science and engineering, and she regularly gives talks in schools and judges entries to the annual Big Bang Fair competition. In recognition of this work, in 2020 Maggie became the first black woman to receive a gold Lord Kelvin Medal. As well as one of the presenters of BBC's *Sky at Night*, Maggie is also the current chancellor of the University of Leicester.

Equality, diversity and inclusivity matter, and many organisations are supporting this. Choose a company you are interested in (for their products, service or possibly a career) and have a look at their website. Find out what they are doing to help all of their workforce feel included and valued.

You may have an image of what someone who works in STEM looks like and you might think that this person isn't like you. There will be someone like you who has a career that you are interested in. Find your role model by asking family, friends, teachers and a careers adviser for help. Explore websites and social media in the areas you are interested in. There are organisations that profile role models and can even arrange professional mentors for young people early in their STEM career. Meeting a person who is like you and has a career in an area of STEM you are interested in can be a game-changer!

Jobs at all levels

It doesn't matter what your academic ability is. If you are interested in working in an area of science, technology, engineering or maths there will be a job you can do. It's worth pointing out that you'll need to do your best in Maths and English as these are often required at Grade 4/C or above (in Scotland – National 4/working towards National 5).

There are different entry levels into a variety of STEM jobs from science, engineering and construction to sport, healthcare and IT. For most people in the UK, you'll need to finish your GCSEs before you can start an apprenticeship. However, in Scotland there are a few Foundation Apprenticeships that can start alongside your National 4/5 studies. Some young people decide to stay in school/college while they are 16–18 and gain some more qualifications and then go on to do a STEM job when they leave full-time education. Often these jobs come with training, or they may be an apprenticeship and may lead to professional qualifications achieved while in employment. Some students choose to stay in full-time education and go to university. Others may find that they try a few things before they find what they really want to do. In Chapter 3, read Josh's career story about becoming an apprentice mechanical engineer during his twenties.

Chapters 6, 7 and 8 (making decisions about study and work) will explore all of this further. But for now, don't assume that a career may be off-limits to you because of the entry level. For example, degree apprenticeships now exist for becoming a doctor. This means that some of the financial burdens of long university courses can be avoided for a young person with the required A levels/Advanced Higher qualifications.

STEAM

Some people argue that the acronym STEM should be replaced with STEAM, where the 'A' represents art.

There are a number of reasons why the arts are important. Let's think about school/college first. Being able to study art, music, drama, dance and other performing arts gives people the space to experiment with their creativity and build confidence in performing and expressing their ideas. These are incredibly important skills that all employer values in the workplace. You may have a STEM career that never involves drama or music, but the confidence you gained when performing may help you to deliver presentations. Product designers need to understand aesthetics and ergonomics when creating new objects, such as furniture and clothing. Engineers need good creative skills in order to problem-solve. Creativity is a highly valued skill that will be discussed further in Chapter 5.

Creative industries cover a broad range of areas such as film, TV, gaming, architecture, fashion, music, marketing and journalism. Many careers in these areas will need good science, technology, engineering and maths skills. New technologies are driving changes within creative industries, leading to more jobs and economic growth. In Chapter 3 there will be a deeper dive into creative industries.

How is this relevant to you? You may be looking for inspiration for possible careers. A person's hobbies and interests could lead to a rewarding career. So, if your passion lies in music, gaming, social media, history or photography, for example, be open to exploring the career opportunities within these areas. If you're good at maths and computer science find out about the career pathways in the creative industries that involve technical skills such as data analysis. How else do you think Spotify is able to choose your daily mix?

Read Molly's story at the end of this chapter about her studies and career in media production.

Charity organisations

Many people are motivated in their career by the causes they are passionate about, for example, animal welfare, ending child poverty and protecting the environment. Charity organisations need STEM talent in order to achieve their goal.

The charity sector is sometimes referred to as the third sector. These organisations want to make a difference to society and making a profit is not an objective. The values that a charity has may be similar to your own. You can find out about the values a charity has (or any company) in their mission or vision statement on their website. Chapter 5 will discuss why it is important to work for a company that has similar values to your own.

If there is an area of STEM you are good at and a charity cause you are interested in, there may be a career opportunity that links the two! Scientists are needed to push medical research. Mathematicians are required in aid agencies to work on the logistics of getting support to those who need it. Engineers are fundamental in designing better equipment, such as developing prosthetic limbs. People with good technical skills are vital for social media and fundraising campaigns.

As your professional skills develop, you may be able to offer help to a charity. For example, the Scottish Tech Army (www.scottishtecharmy.org/) volunteers time and skills to help third sector organisations solve problems with their technology and business systems.

Read Catriona's career story at the end of Chapter 4 to find out how she uses her technology knowledge to help a local charity.

Non-STEM roles in STEM organisations

There is probably a big STEM employer close to where you live. They will employ lots of people and some will not have a traditional STEM job. There may be staff who specialise in company law, human resources, business support, administration and marketing. All of these roles are vital in order for the company to function efficiently. For example, the marketing team promotes the company's services or products, and the human resources (HR) department oversees staff wellbeing and career development.

If you are employed within a STEM company you will be working with colleagues from different backgrounds and specialisms. You will all work more collaboratively if you understand everyone else's role and how this assists you with your job.

You may have friends who don't enjoy science, maths and tech subjects at school and think they have no interest in STEM. It is important for them to know that they could have a really great career within a STEM company.

STEM skills required in all jobs

Whatever job someone has, they probably use STEM skills. This is true of our day-to-day lives, too. Making good decisions about looking after our health involves science. Using our phones requires tech skills. Doing DIY can involve simple engineering and planning a holiday uses our maths skills.

Knowing how to use mechanical and electrical equipment required for a job involves STEM skills. Being able to handle money correctly involves numeracy skills. The hospitality, sport and beauty sectors also use science knowledge as well as tech and numeracy skills.

'Digital literacy' is a phrase you may have come across. In today's society it is difficult to avoid tech. Many things are done online, from controlling our personal finance to applying for a job. Being able to navigate a website, use smart devices in our homes and use apps on our phone all make life easier.

Some jobs may not appear to involve STEM; however, they require good technological and digital skills. For example, a marketing assistant may run the social media for the company. In some jobs you may be expected to use and maintain databases. A database is a collection of data (information) that can be added to, searched and presented in an organised way. Spotify, Facebook and TikTok all use databases to store and retrieve information in an efficient way.

Chapter 5 will explore how numeracy and technical skills (including digital literacy) are important in the world of work in greater detail.

Green: Skills, jobs, careers and economy

You may come across terms such as 'green skills', 'green jobs' and 'green careers'. Closely associated are the phrases 'green economy' and 'net zero'.

'Net zero' means that the quantity of greenhouse gases released into the environment is the same as the quantity removed. The UK government is striving for all sectors of the economy to be 'green' that is, net zero, by 2050.

The National Careers Service website advises that any career that helps the UK work towards net zero could be considered a green career, and the skills required to work in these jobs are referred to as green skills (https://nationalcareers.service.gov.uk/careers-advice/green-careers).

Fascinating fact

Microbiologists and nutritionists are discovering the importance of our gut microbiome. The foods you eat can affect your gut microbes, and they in turn can affect your health and wellbeing. Eating more plants (veg, fruit, nuts and seeds) and fermented foods (such as natural yogurt, kefir and kimchi) and less ultra-processed food is better for our health.

Conclusion

Everyone needs to know that a STEM career can be for them. There is a potential loss of talent if people think STEM isn't for them. The world's problems can be solved, and society can be made better for all if a wider range of people are involved in the organisations that can make a difference.

Perhaps there are areas of science and technology that you are interested in but not sure how this could lead you to a career. Be curious – by finding out a little more about the sector or companies that you use, you may find out about a job you never knew existed.

You may be interested in a career in STEM but need to find a role model with a story similar to yours. Try to find a role model that you can read about or, better still, talk to.

Keep an open mind about STEM and the world of work. There are many jobs within the STEM sector that are not directly STEM-related but are very important in helping the company to function. The skills we acquire that are related to STEM are useful in our day-to-day lives as well as in the workplace, even if a person thinks they are not interested in STEM. Digital literacy is an important example of one of these skills.

 REFLECTION ABOUT CHAPTER 2

Use your reflection notes from Chapter 1 to think in more detail about why you are interested in science, technology, engineering and maths.

Is it because you like/use a product or service (e.g. medicine, cars, gaming, fashion, music)?

- How do you use this product/service?
- What can you find out about the history and development of this product or service?

Are you passionate about a cause and want to find out how you can help to make a difference (e.g. improve access to resources such as food and water, tackle environmental problems, improve health and wellbeing)?

- What are the problems in your area of interest that need to be solved?
- How can STEM help to solve these problems?
- How would you like to be involved in this (working on the problem, communicating with the public about the problem/ solution)?

Keep a note of your thoughts as they'll be useful in Chapters 3 and 4 when you'll find out more about sectors and jobs.

Career story: Molly

**Multi-Platform Podcast Producer,
Bauer Media UK, London**

Describe your job

I am a podcast producer for Bauer Media UK. This is a massive brand that's involved in radio and magazines, such as Kiss FM and Heat. I am part of the new podcast team, and I work on two titles.

At Bauer, we are developing the social media presence of the podcast to create the brand. This includes creating video content to go with the podcast.

One of the podcasts I work on is called Mother Half. I do the guest bookings, research the guest and work with their agents to agree on what will be discussed, for example, if they have a new book coming out. I sort out the running order for the day, get the kit ready and set up for the day. I'm responsible for recording the podcast and then doing the editing. We then contact the press and media outlets to promote the podcast.

Every day is different and it is an exciting, fun area to work in. There are lots of creative minds around you and nothing is too over the top when you are coming up with new ideas. The challenges are that creative people aren't always very organised, and there can be some 'picking up the pieces' to do! There can also be a clashing of egos in this industry! It's non-stop, and there can be early mornings and working overtime.

What were your interests when you were a teenager?

I was really arty as well as interested in history and reading. Science and maths weren't really for me. At college I did A levels in history, archaeology, English literature and media studies. I enjoyed the sound side of media although we studied video, too.

My brother and I enjoyed using tech to make our own silly videos, trying to create special effects. This gave me a good base of knowledge for learning how to use technology when I was studying production at university.

Did you know what you wanted to do after your A levels?

Not really, no. It's such a big step, but I knew I wanted to go to uni. I absolutely loved archaeology and was torn about what pathway to do next, media or archaeology. I looked into what you'd do in the two careers and the progression and thought was more exciting to get involved in broadcasting. I looked into which universities did the best media courses and visited my top two places. One was a modern city and I knew it didn't have the right vibe for me. As soon as I stepped off the train in Lincoln, I knew that was the city for me. It has a mixture of old and new, lots of history and historic buildings.

The main reason for me picking that course was that you got to study a bit of everything in the first year, such as radio, film, TV, visual effects and CGI, script writing and photography. I got to learn how to use special effects software, honed my radio skills on the university and local radio channels, and shopped in charity shops to create set displays for making videos.

What did you do after your degree?

After the degree I went straight into a master's degree in digital media at the same university. I wanted to stay in Lincoln because I had friends, and my boyfriend there had a part-time job to help pay for rent. It was a very theory-based master's that I didn't expect. I might have done a different master's degree if I had realised exactly what it involved. However, the digital knowledge I learnt helps me now as I work across multi-platforms.

When I finished the master's, I thought 'Agh . . . I have to do something' and I started applying everywhere for a job. I found working for a kid's radio station but it was in London and I didn't really know if I wanted to go there. But I put myself 'out there', ended up getting the job and was thrown into London and have loved it ever since.

How did your part-time jobs help you?

During college and university, I worked in lots of hospitality roles, including the local chip shop, supermarket and coffee shops. This threw me into the world of work and pushed me out of my comfort zone, which was a good thing as it helped me learn about people. Having a part-time, customer-facing job can improve your confidence in social situations. This helps me now when I work with guests on the podcast and I go into 'barista-mode' making sure they have everything they need.

3 STEM sectors (making things)

Introduction

The term 'sector' is used for a broad group of industries that all have similar aims and goals. The job website Indeed explains that sectors can be split into four groups:

- extracting raw materials (e.g. mining and agriculture)
- transforming raw materials into products (e.g manufacturing and construction)
- providing a service (e.g. hospitality, healthcare and education) and
- creating new knowledge (e.g. information technology and research that takes place in higher education, also called academia).

The four groups of sectors all contain STEM and work together to develop the economy and benefit society. An example of this is healthcare: research takes place in universities leading to the development of new medicines that need to be manufactured. Another example is hospitality: plants and meat are farmed and then used to manufacture the food, developed by food technologists, that is served in your favourite restaurant.

For the purpose of this book, 12 STEM sectors have been selected and split into two groups, 'making things' (discussed in this chapter) and 'helping others' (discussed in Chapter 4). Each sector will include five examples of STEM jobs that are in demand. Other sources, such as websites and books, may have slightly different lists.

Grouping the STEM sectors

STEM sectors (making things)

1. Mineral products
2. Agriculture and food
3. Construction
4. Manufacturing
5. Life sciences and Pharmaceuticals
6. Creative industries

STEM sectors (helping others)

7. Energy
8. Conservation
9. Finance
10. Health
11. Logistics and transport
12. Information technology

These two chapters should give you a good starting point for researching areas that you are interested in. Also, read the brief description of sectors you're not interested in, as you may discover a new career idea! (The A–Z at the end of the book has another 60 career ideas!)

Trying to classify each industry into one sector can be difficult. Pladis Global is the company that makes Jaffa Cakes. They operate from 'field to kitchen', so do we put them under agriculture, manufacturing, food technology, logistics or retail? It doesn't really matter, as long as you appreciate the breadth of what they do and what opportunities exist in a STEM company such as this.

> **MORE INFORMATION ABOUT . . .**
>
> *UK Armed Forces careers*
>
> It is difficult to put the military in a sector. The roles are diverse, ranging from cooks, engineers and medics to soldiers, helicopter crew and intelligence specialists.
>
> The British Armed Forces (British Army, Royal Air Force and Royal Navy) is one of the biggest employers in the UK. As well as combat, their functions include peacekeeping and humanitarian aid.
>
> There are lots of opportunities for training and career progression, and scholarships are available to become an officer.
>
> Visit the National Careers Service and the Ministry of Defence websites for more information.
>
> https://nationalcareers.service.gov.uk/careers-advice/careers-in-the-armed-forces
>
> www.gov.uk/government/organisations/ministry-of-defence

3 STEM sectors (making things)

Six STEM sectors involved in making things

Mineral products

Did you brush your teeth this morning, have a drink out of a cup or glass and use your phone? Then you have used products made from the raw materials that mineral industries extract from the earth. The mineral products sector is one of the largest manufacturing sectors in the UK. It provides raw materials that are essential to our contemporary lifestyle. This sector is forward-thinking and progressive, providing critical materials in a sustainable way that make a positive contribution to society.

The materials quarried and mined are needed for a variety of other sectors such as construction, agriculture, pharmaceuticals and technology. The sector is at the forefront of providing solutions to the global challenges we face. An example of this is the collaborations between industry and academia to produce net-zero cement for the construction industry. The components in new technologies contain minerals extracted from the earth, such as silicon in solar panels and lithium used in rechargeable batteries, both mined in Cornwall.

Examples of five important careers within the mineral products sector you may be interested in are:

- **Quantity surveyor** – Are you good with numbers and like to get value for money? This role involves minimising the costs of the project (e.g extracting and processing a mineral ore) in the safest and most efficient way.
- **Mechanical engineer** – Do you like fixing things? This role involves developing, testing and building machinery. A mechanical engineer will work closely with other staff in the business ensuring budgets are met and risks are managed.
- **Health and safety manager** – Do you pay attention to detail and care about everyone working safely? This role involves using legislation and policies to promote a positive, healthy working environment for all employees and site visitors.
- **Geoscientist** – Are you interested in learning more about earth science at university? Geoscientists are needed in the mineral products sector to help discover reserves of raw materials and help to make their extraction commercially viable.
- **Environmental manager** – Do you care about the environment and have good leadership skills? This job involves helping the company to work more sustainably, reduce their negative impact and meet environmental legislation.

Read Josh's story at the end of this chapter and Lauren's story (Chapter 1) to find out more about working in the minerals sector.

Mineral products are required to make just about all of the things we use in modern life. Companies that extract raw materials want to do this in a cost-efficient way that minimises their impact on the environment and improves the habitats in the areas they work. As part of the life cycle of a quarry there are collaborations to help local communities learn about the geology and industry. Companies also work with environmental charities to ensure the habitats and biodiversity are restored, or even improved, after the extraction process has finished.

Working well as part of a team and collaborating with partner organisations is important in this sector. Mining and quarrying are global activities, and many companies operating in the UK will provide opportunities to work in other countries.

For more information, visit the Minerals Matter website (https://minerals-matter.co.uk/).

Agriculture and food

This sector covers everything from 'farm to fork'. This is much more than just muck and manual work as there are a huge range of industries involved in getting food to our plates.

- Farming – this is becoming more technical and scientific as farmers try to maximise the yield from plant crops and animals using technologies such as GPS and drones.
- Research – universities and colleges work with farmers to develop new breeds, crop varieties and farming and fishery techniques.
- Manufacturing food – covering everything from creating new foods, making products more healthy to running efficient production lines.
- Packaging and transport – designing new wrappers to keep food fresher for longer and ensuring food from farms and factories gets to the shops efficiently.
- Retail – wholesale to shops and hospitality, and shops selling to the public.

The agri-food sector is an exciting area to work in, involving looking after our countryside, improving animal welfare, innovating healthier products and keeping costs down for the public.

Examples of five careers within the agri-food sector you may not know much about are:

3 STEM sectors (making things)

- **Animal geneticist** – Do you enjoy learning about how genes and DNA affect an organism's characteristics? This job uses population genetics and breeding programmes to develop desired traits in farm animals and improve their welfare.
- **Soil and plant scientists** – Are you interested in combining environmental chemistry and biology to help secure food production? This role involves monitoring and improving soil quality, and finding ways to conserve water and improve crop yield.
- **Agricultural economist** – Are you good with numbers and interested in business? Using financial management, agricultural economists monitor the market trends and policy requirements to maximise the industry's profit.
- **Food scientist** – Are you good at problem-solving and have attention to detail? This job uses a range of STEM disciplines to improve food processing, including controlling nutritional levels and finding ways to preserve food for longer.
- **Product developer** – Are you passionate about the quality of manufactured food? A product developer improves existing foods and develops new ones to meet consumer demand and government health guidelines by taking a scientific approach.

The pandemic, environmental concerns and the effect of conflicts in Europe have all shown that food security is very important. The agri-food sector is not one of the biggest in the UK, but there are good career opportunities involving all of the STEM subjects. For inspiration, find out about where the food you like to eat comes from and how it was developed.

For more information, visit the Tasty Careers and Lantra websites.
https://tastycareers.org.uk/
www.lantra.co.uk/careers

Manufacturing

This sector is huge and covers everything from making cars to making dinner plates! The UK has a proud history of making things and is one of the world's leaders in innovating and manufacturing. There will be businesses near you involved in manufacturing, and some may be global names you have heard of and some could be SMEs that you may not have paid much attention to. (See box below for more information about SMEs.)

Manufacturing industries will have factories where there are production lines. They may also be involved with the research and development of products and technologies. These industries may be making things for other sectors, such as equipment for the construction industry or devices for the IT sector. They may be manufacturing products that will sell on the high street.

This is such a huge and varied sector that it is difficult to pick out representative careers, but here are five that you might be interested in finding out more about:

- **CAD engineer** – Do you have a good understanding of how science, engineering and maths principles are used in the design of functional objects such as furniture or vehicles? A CAD (computer-aided design) engineer uses software to plan how products can be designed and made.
- **CNC engineer** – Do you enjoy learning computer programming? Computer numerical control (CNC) engineers use programming to automate a machine's operating. The demand for CNC engineers will increase as more industries use automated processes.
- **Production welder** – Are you a practical person who enjoys working with metal? Welders are skilled professionals who use different techniques with precision to join metal pieces together to make metal products. They can be employed in a variety of industries, from automotive and aerospace to construction.
- **Quality control technician** – Do you have attention to detail? In this role, products are tested to ensure they meet national and international standards and are fault-free. This is important for health and safety and maintaining the company's reputation.
- **Maintenance engineer** – Are you a practical person who enjoys learning how things work? Maintenance engineers are able to adapt and problem-solve to keep the production line running and help the business run economically.

This is a huge sector and exploring the websites of the companies that make the products you like to use is a good way to find out more. How do you know if a career in manufacturing is for you? Your interests may lead you to this career, such as an interest in being creative and making things, or something specific, such as cars. You may see local training opportunities available to you that are advertised on job websites and seem interesting.

Read Ethan's story, at the end of this chapter, or Dionne's story (Chapter 10) about their roles in manufacturing aeroplanes and cars.

For more information, visit Make UK website (www.makeuk.org/).

3 STEM sectors (making things)

> **MORE INFORMATION ABOUT . . .**
>
> *Small to medium enterprises (SMEs)*
>
> SMEs are businesses that employ fewer than 250 people and turnover less than about £43 million.
>
> So, how do you learn more about your local SMEs?
>
> - Search on a reputable job website, such as *Indeed*, *Reed* or *Monster*, for jobs within a 20-mile radius of where you live.
> - Look at the names of businesses that you pass regularly on the way to school and college.
> - Ask friends and local family where they work.
> - When you find out the names of some local SMEs, do some research about the sector they operate in and what they do.
>
> Another way to find out about a region's SMEs and local industry is to research on the region's Local Authority website. Search terms could include:
>
> - priority sectors
> - business development strategy
> - regeneration
>
> ### Why is this important?
>
> Over 99% of all UK businesses are classed as SMEs, and they provide approximately 50% of the private sector employment. So, there is a good chance that you may one day work for an SME, and it might be a company that you don't know much about.
>
> Knowing about the employment opportunities in the area you would like to live in is part of developing your career management skills.

Construction

This is so much more than muddy boots, bricks and building roads. You might not think you are interested in working in construction, but this industry is part of your day-to-day life. Do you look around your built environment and wonder if it could be better and safer? Are you concerned about how we can have sustainable travel that suits everyone's needs? Then a career in construction could be for you.

This sector is working hard to improve its image, and many companies are actively seeking a diverse workforce. There are many roles in construction, not all based on building sites, and a variety of pathways into careers. It is often a good idea to gain work experience. This helps you find out a bit more about the industry by speaking to people doing the jobs. (See Chapter 10 for more advice about work experience.)

Here are five examples to show the breadth of entry levels and careers in construction:

- **Architect** – Would you like to use your creativity and skills to design new structures or renovate old ones? A degree or degree apprenticeship is the usual route to become an architect, although you could start as an assistant and then obtain the qualifications.
- **Civil engineer** – Can you work well in a team and do you have good computer skills? In this role you use modelling software and the designs from the architect to plan construction projects for site managers and contractors to follow.
- **Site manager** – Would you like a leadership role on a construction site? Site managers ensure the work is done safely and within the timeframe and budget. The career pathway could be a college course, apprenticeship or degree, but you'll also need experience of being on site and assisting with managing projects.
- **Electrician** – Would you like a practical job that applies your knowledge of electricity and circuits? Qualified electricians can be self-employed or employed by a company. They install, inspect and test electrical equipment in a variety of building projects.
- **General construction operative** – Would you like to work on a building site but not sure about what training to do? Try to get some work experience and keep physically fit. Once employed as a general worker, there are many training routes that lead to formal qualifications such as joiner, bricklayer and road surfacing.

As society requires more housing, new roads and other transport links, the need for building responsibly and being sustainable is more important than ever. If you care about people living in well-designed and affordable homes, then a career in construction could be for you.

For more information, visit the Go Construct website (www.goconstruct.org/).

Life sciences and pharmaceuticals

This sector could be included in manufacturing, but, as the pandemic showed us, pharmaceuticals and medical supplies are so important that this sector deserves its own section in the book.

3 STEM sectors (making things)

Life sciences and pharmaceuticals is a high-tech sector that involves medical technology (devices and diagnostic equipment such as blood pressure monitors) and pharmaceuticals (the research, development and manufacturing of medicines). Companies involved in this sector work in partnership with academia (where much of the research takes place) and the NHS (the main purchaser of biotech equipment and medicines in the UK).

Five careers within the pharmaceutical/life science sector are:

- **Research scientist** – Do you enjoy biology and chemistry, pay attention to detail and have good problem-solving skills? A degree in a subject such as biomedical science can lead to a career as a research scientist. They use experiments to help find how a drug may be used as a new medicine.
- **Chemical engineer** – Would you like a career that uses science and maths knowledge? Chemical engineers research how to make the best use of a substance, for example, the most efficient way to scale up manufacture of a drug. They work closely with research scientists and production managers.
- **Biomedical engineer** – Do you enjoy all areas of STEM and want to help make life better for people? This job combines engineering, computer skills and medicine to design and make medical equipment, such as replacement joints, artificial limbs, robotic equipment to support surgery and diagnostic devices such as blood glucose monitors.
- **Production manager** – Once you have some experience in manufacturing or pharmaceuticals, would you like a leadership role? In this job you'll supervise the production team, ensuring pharmaceutical products are manufactured safely, on time, within budget and meet legal quality standards.
- **Supply chain planner** – Are you able to 'see the bigger picture' and can you work well in a team? Supply chain planners use software to assist with forecasting demand, managing stock and analysing data. Many companies offer apprenticeships in this role that could lead to further study.

New medicines are required to combat disease, and new technology is pushing the frontiers of medical devices – for example, antiviral drugs to treat Covid-19 and insulin pumps that help to control diabetes. Many pathways into careers within the life science/pharmaceutical sector involve work experience and industry placements. (See Chapter 6 for more information about industry placements.)

Read Hannah's story in Chapter 7 for information about studying pharmacology.

For more information, visit the Association of the British Pharmaceutical Industry at www.abpi.org.uk/careers/ and Cogent Skills at https://cogentskills.com/.

 HERO IN STEM: PROFESSOR DAME SARAH GILBERT

Sarah Gilbert is a professor of vaccinology at the University of Oxford. She studied biological sciences at the University of East Anglia and completed her PhD in biochemistry at the University of Hull. Her early career focused on genetics and malaria research. In 2005, Professor Gilbert joined the Nuffield Department of Medicine (based at the University of Oxford), and her main area of research became vaccine development.

Right at the start of the Covid-19 outbreak, in January 2020, Professor Gilbert and her team at the University of Oxford were able to apply their knowledge from previous work on flu vaccine development in the fight against the new coronavirus. Working in partnership with AstraZeneca, she became the lead scientist on the search for a vaccine and by April 2020 the clinical trials in humans began. Other collaborations were also in the race to develop a vaccine and in January 2024 the World Health Organization estimated that over 1.4 million lives had been saved in Europe due to vaccination against Covid-19.

In 2021, Sarah Gilbert was made a Dame in recognition of the work she did for the safe and effective development of the Oxford AstraZeneca vaccine. Global recognition of her achievements also came in the production of a Barbie doll made in her image.

Creative industries

This is a huge sector that involves a range of industries, from marketing to fashion and TV to gaming. One way to understand the breadth of creative industries is to put them under four headings:

1. the screen sector: including film, TV, games, animation and visual effects (VFX)
2. the arts: including art and craft, culture and heritage, and theatre
3. pop culture: including fashion and music
4. media and communications: including marketing, journalism, digital media and radio.

3 STEM sectors (making things)

Technology underpins all of the creative industries, whether that is having good digital skills to enable you to produce the soundtrack for a film or having professional technical skills so that you can create, for example, new games using code. In order to continue flourishing and contributing to the UK's economy, the creative industries need technicians, scientists, computer programmers, data analysts and those with good digital skills (see more about digital skills in Chapter 5).

Five examples of careers in the creative sector that you may not have thought about include:

- **Chief lighting technician (gaffer)** – Are you creative, and understand the physics of light and how electrical circuits work? This job uses lighting to bring a film or programme to life, as envisaged by the director of photography. The chief lighting gaffer supervises their team, keeping everyone safe.
- **Data technician** – Do you have excellent computer skills? Data technicians can provide technical support within VFX companies, organising and transferring data, and storing computer video and animation files.
- **Games animator** – Do you understand art and computer gameplay? Games animators use programming software to bring the artist's creations to life when new games are created. They are able to work in an organised way to manage their data files and meet deadlines.
- **Fashion trend analyst** – Do you enjoy popular culture and fashion, and have good numeracy and communication skills? These analysts predict fashion trends to help the company maximise profits with minimal risks. They conduct market research and interpret data about sales, emerging trends, sustainability and household finances to predict what fashions will sell next season.
- **Marketing manager** – Would you like a career that uses a mixture of psychology, technology and business knowledge? Marketing managers promote the organisation's product or service and for charities, their cause. They need good IT, social media and financial skills as well as excellent interpersonal skills.

Perhaps you enjoy TV and film, music, fashion or gaming but are unsure how you can have a career in one of these areas. We sometimes ask our family and friends for advice about career pathways. If they don't know much about your career interests, they may feel unable to support you. That is why it is important for you to know where to turn for reliable information. Your school/college careers adviser can help, as can reputable websites like Discover Creative Careers (https://discovercreative.careers/).

(Remember, 'creative industries' and 'creative skills' are different things. See Chapter 5 for more information about creativity as a skill.)

In Chapter 2, read Molly's story about her career as a podcast producer.

For more information about careers in the creative industry, visit the following websites:

Screen Skills, www.screenskills.com/
Meet Eric, www.meet-eric.com/creative-industries
Into Games, https://intogames.org/

 Fascinating fact

Two physicists at the University of Manchester discovered the carbon macromolecule *graphene* in 2004. They played with sellotape and graphite and eventually managed to get a layer one atom thick! Graphene has amazing properties: it's the most conductive material ever made, is flexible and stronger than steel. The uses for graphene are still being discovered, including applications in medicine and communications.

Conclusion

Remember that each sector will contain many industries. For example, the manufacturing sector includes automotive (making vehicles), aerospace (e.g. making satellites), plastics and semiconductors. (Semiconductors are electronic components that are needed in equipment such as smart phones, TVs, computers, cars and medical equipment. Businesses that make semiconductors are suppliers to businesses within many other manufacturing industries and other sectors.)

Categorising things isn't always useful, so keep an open mind. If you enjoy using console games, perhaps you could find out about how drone technology is useful in many sectors. If you are thinking of doing a biomedical sciences degree, career opportunities could be available in pharmaceutical manufacturing, health and safety services, medical research, as well as healthcare.

 REFLECTION ABOUT CHAPTER 3 CAN BE FOUND AT THE END OF CHAPTER 4.

Career story: Ethan

Engineering Degree Apprentice, Airbus

Why did you want to become an engineer?

I've always been interested in aircrafts and wanted to work within the aviation industry as a pilot or an engineer. I moved to the UK from India when I was 16. We lived close to Heathrow Airport under the flight path so I would see aircraft flying. British Airways did lots of events and workshops that I attended, and I got to board one of the aircraft that was in for repairs.

During sixth form, I took part in a Transport for London (TfL) competition for all the schools, and my team were finalists. Part of the prize was two weeks of work experience. We saw all aspects of the engineering facilities within TfL in the first week, including the control room, and the second week was with the emergency response unit for TfL. Working with them, I realised how engineering could be applied within the aerospace sector. This work experience was so helpful when I was applying for the degree apprenticeship as I could talk about it in the interview.

How did you decide what A levels to study?

I took maths and physics because I knew they were a requirement for studying engineering at university. I also took A level biology and started chemistry but didn't enjoy it.

Why did you choose to do a degree apprenticeship?

When it came to UCAS application time, I talked with the careers adviser and my headteacher. They knew I wanted to do aerospace engineering and recommended I go to the events at Heathrow Airport. We also had a talk about apprenticeships, and I was encouraged to look into this as an alternative way to get a degree. The end qualification is the same as doing a three-year degree at university, but 'on the job' experience is really valuable, and this is a debt-free option.

It's not an easier option as it can be hard to get onto a degree apprenticeship. Studying and working is difficult and the company wants to know that you can handle it. I applied for a degree apprenticeship with Airbus. I was invited to attend an assessment day that involved an interview and team tasks. They want to see how you are as a person, how you solve a problem and work within a group. Airbus was my dream company to work for, so getting an offer from them was amazing.

Describe your apprenticeship

It can be a challenge to manage studying and the day-to-day job, and it's a fine balance to juggle everything. Good time management is really important. We spent the first year at university learning engineering principles. At the end of this year, the managers and apprentices of the different areas gave us a talk about what their division does and the opportunities available. There are a lot of different areas you can do placements in, including structures, fuel systems, landing gear and testing. I made a list of what I was interested in, which was mainly design and research and testing (R&T).

During the rest of the degree apprenticeship we've done rotations that tend to last three to six months. I am in my final year, attending university just one day a week, and am currently in my dissertation placement. I've had placements within the wings structures area and the wing design office which is involved in continuous product development and testing. I have also spent some time within the regulatory side of the business and the airworthiness certification teams. During the degree you gain other qualifications including level 2 (Aerospace and Aviation Engineering) and level 4 (Advanced Manufacturing).

I enjoy working with so many people who have some amazing stories to share about the projects they have worked on. At Airbus, there is a focus on 'early careers' and knowledge-sharing, so the team is willing to help you learn.

What would you like to do when you complete the apprenticeship?

I would like to stay within the aerospace sector, and there are many opportunities in Airbus, based on what you are interested in. Airbus designs and manufactures commercial and military aircraft, including helicopters, and space satellites. I would like to work in the R&T division as a design engineer and gain professional accreditation including Incorporated Engineer (IEng) and Chartered status.

Career story: Josh

Mechanical Engineer Apprentice, Imerys, South West England

What is your current job?

I am due to complete my four-year engineering apprenticeship with Imerys, achieving a level 3 Engineering Technician (Mechatronics) Apprenticeship.

I currently work in Cornwall in a processing plant, which dries and secondary processes china clay.* I work with compressors, conveyors, pumps, motors and gearboxes. I am also cross-skilled and able to carry out electrical work. My day-to-day work is varied. I can be carrying out routine maintenance work (proactive maintenance) or repairing faulty equipment (reactive maintenance). I like to learn about how the plant operates so that I know how each part of the process is linked and how important it is to the business that it operates effectively and efficiently.

I joined Imerys when I was 27 and I'm one of the oldest apprentices in my cohort. There's a mixture of school-leavers and people in their twenties. I've enjoyed this apprenticeship, spending the first year full time in college, learning the theory and practising hand skills. In Years 2–4, we've spent more time in the business working with the skilled engineers, getting to know the processes and machinery while attending college one day a week. It was good to interact with other people and use the skills learnt in college.

In the future I would love to go further, maybe into management if I can. If the opportunity to do a HNC or HND is offered, I will take it.

(*China clay, or kaolin as it is also known, has been mined in mid-Cornwall since the 18th century. It is sold worldwide and transformed into sophisticated engineered minerals which are prized for high-end ceramics, paints, adhesives, plastics, rubber, cosmetics, pharmaceuticals, paper and cardboard.)

What did you do between school and starting the apprenticeship?

At school I was very sporty and I liked IT, too. But I didn't do either at college because while trying to decide what to do, I left it too late to apply. So I ended up doing BTEC level 1 Construction Studies and BTEC level 2 Plumbing. Even at that age I knew that plumbing and construction skills would be useful for my future.

I worked in various retail jobs from 18 until I was 26 and lost my job due to Covid. I spent a few months looking for work when I found this mechanical engineering apprenticeship. I wrongly thought I had to be between 16 and 24, but I took the chance, got an interview, did the various tests and got the job.

What do you enjoy most about being a mechanical engineer?

Being able to make something from scratch, such as a bracket for fixing something to a wall! Sourcing the materials, measuring it myself, knowing that I put all my effort into it and that it can now be used. It's having the skills and thought-processes to fabricate something, what does it need to do? Where does it need to go?

What are the challenges in your job as an engineer?

There are certain times when I am 'tested': Do I have the knowledge to do a task, or do I need to seek help? I can be in a situation and I have to think, do I completely understand this or do I need that little bit of backup. Just to be safe, it's important to seek assistance from others.

What advice would you give your 16-year-old self?

Work hard and really consider an apprenticeship. It is a great way to learn the skills, get the industrial experience and earn while you learn. If someone tells you no, work hard and prove them wrong. It's going to be hard, work at it and think about what can come afterwards.

4 STEM sectors (helping others)

Introduction

Six sectors involved with 'making things' were introduced in Chapter 3. This chapter will focus on six sectors involved in 'helping others'.

It can be restrictive to group sectors under individual STEM headings. You need to understand the importance of collaboration between disciplines. Here is an example: A software engineer who uses code to write applications could work in any of the sectors listed in Chapters 3 and 4. But it would be useful if that person knows a bit about the sector they are working in. IT specialists working in health understand how their work fits into the aims of the healthcare provider, as well as the needs of the software users.

Six STEM sectors involved in helping others

Energy

The energy sector is involved in delivering electricity, gas and other hydrocarbons, such as oil, to all areas of society including homes, schools and businesses. There are many different industries within this sector that can be grouped as:

- generating electricity mainly from gas, wind, nuclear energy, biomass.
- balancing supply and demand of electricity through the National Grid. New technologies will help to achieve this as the country increases its use of electricity for transport and heating homes.
- supplying customers with gas and electricity. Energy retailers buy the gas and electricity and sell it on to homes and businesses. This is a competitive market as consumers can choose their suppliers.
- regulating the gas and electricity markets. The Gas and Electricity Markets Authority issues company licenses to operate, promote competition and protect consumers' interests.

The sector is undergoing huge changes to ensure reliable and secure energy supplies while cutting carbon emissions.

The government and international organisations are investing in UK clean energy technologies, and this means lots of opportunities for exciting careers in a range of roles including technical, financial and strategical.

Here are examples of five careers that demonstrate the opportunities:

- **Electrical engineer** – Are you able to analyse and pay attention to detail, and do you enjoy maths and physics? An electrical engineering qualification can lead to opportunities in projects involved with wind farms, solar farms and hydroelectric schemes.
- **Renewable energy analyst** – Would you like to use your numerical skills to improve the efficiency of renewable energy sources? These analysts work with data to monitor the day-to-day performance of electricity production sites and report the information to the wider team.
- **Power infrastructure engineer** – Would you like an engineering role that involves improving the country's electricity network? These engineers work in a project design team within a company that specialises in transporting electricity from where it is generated to where it is used.
- **Smart meter engineer** – Enjoy working with electrical components? Vocational qualifications from college and on-the-job training in installing gas and electricity meters are required. They specialise in how to install and commission smart meters.
- **Home energy adviser** – Do you have excellent communication skills and enjoy helping people? This role involves advising homeowners about how to make their house more environmentally friendly and cheaper to run. They keep their knowledge of technological developments with smart appliances and home insulation up to date.

It's an exciting time to work in the energy sector. Research and innovation are required to help the UK reach the net zero target set for 2050. More of us will use smart appliances to help households use electricity more efficiently. Electric vehicles and heat pumps in our homes may one day be available for everyone as the development of the technologies helps prices for consumers to fall.

Read Alex's career story in Chapter 9 to find out how he uses his maths PhD to help in the quest for clean energy.

For more information, visit the Energy UK website (www.energy-uk.org.uk/).

Conservation

Conservation is the study and protection of life on earth. This sector is important in protecting species and restoring habitats. Working in conservation involves many different disciplines including biology, geography, economics and politics. Many types of organisations employ conservationists for a number of reasons:

- Local and national governments: need conservationists to work on new policies and encouraging best practices.
- Charities: employ conservationists to look after habitats and raise public awareness of issues, for example, the Woodland Trust.
- Businesses: require conservationists to help reduce the impact of their work and help them to be more sustainable.
- Academia: universities and colleges are where the research and collaboration occur to help develop knowledge and create new ways of solving the threats of habitat loss and climate change.

The conservation sector offers a variety of careers from scientific (e.g. veterinary surgeon or nurse) to technical (e.g. marketing, film-maker) and financial (e.g. fund-raiser, economist).

- **Veterinary nurse** – Would you like to use your love of biology to help care for animals? You'll need to complete a Royal College of Veterinary Surgeons accredited course and can then specialise, for example, in zoo animal care or the rehabilitation and reintroduction of wild animals to their habitat.
- **Ecologist** – Do you care about the environment and have good scientific skills? Ecologists use field surveys to collect information about how human activity is affecting the natural world. Early careers may be specialised, for example marine habitats, with more senior posts involved with legislation and policy.
- **Environmental consultant** – Would you like a career that combines your communication skills and environmental knowledge? Understanding legislation, environmental consultants are involved in research and offering advice on a range of issues such as pollution, flood risk and new renewable energy schemes.
- **Charity fund-raiser** – Are you passionate about the environment and able to think of creative ways to raise money? As well as raising money for the charity, they build relationships with individuals and organisations that may support the cause.
- **Education officer** – Do you have excellent communication skills and enjoy talking to people? Education officers help the general public understand environmental issues and their role in the solution. This role may require a teaching qualification.

Read Ruaridh's story in Chapter 6 about his career in forest conservation.

For more information, visit:

Conservation Careers, www.conservation-careers.com/
Lantra, www.lantra.co.uk/careers

 HERO IN STEM: HAMZA YASSIN

Winning *Strictly Come Dancing* in 2022 enabled Hamza to further explore his creative side and introduced many of us to the wildlife cameraman and presenter.

Hamza was born in Sudan and moved to the UK when he was eight. He studied for a degree in zoology with conservation at Bangor University and achieved a master's degree in biological photography and imaging from the University of Nottingham. This course enables students to develop photographic, technical and journalism skills to capture images of wildlife and communicate this to an audience.

Ornithology is his passion, and Hamza has filmed many of the UK's birds of prey including golden eagles and sparrowhawks. Hamza's filming work has included many national and international projects including the CBeebies series *Let's Go for a Walk* and filming polar bears for the Sky series, *Predators*.

Since he was 21, Hamza has lived in a part of Scotland called the Ardnamurchan Peninsula. He enjoys exploring the local area and taking part in the Highland games.

MORE INFORMATION ABOUT . . .

Sustainability, sustainable careers, careers in sustainability

These are buzzwords at the moment, but let's consider what each term means and how this can affect your career:

- **Sustainability**

Many organisations want to decrease their impact on the environment by using energy more efficiently and reducing waste. They may encourage their employees to follow processes that promote social responsibility and caring for the environment while still allowing the company to achieve its economic goals.

> - **Having a sustainable career**
>
> You will hopefully be working for many years and being proactive about your career can help to make this work fulfilling. Reflecting on your work life and wellbeing, adjusting your work behaviours and taking part in continuous professional development can help you to be ready to make the most of new opportunities when they arise.
>
> - **Careers in sustainability**
>
> This involves working in a sector, or a job role, that focuses on improving the environment or solving global issues. This includes extracting raw materials from the ground, building responsibly, providing reliable energy, and food and water security.

Finance

The UK financial sector is important to the national economy and is a world-leader in providing financial services across the globe. There are many exciting opportunities in corporate banking, insurance and asset management (investments).

There are also career pathways within organisations across all sectors as companies require accountants and other finance staff. With the correct qualifications, experience and entrepreneurial skills, there may be the opportunity to be self-employed, for example providing tax auditing or financial advice.

Five examples of careers in finance that you might not have heard of:

- **Financial adviser** – Do you have excellent communication and numeracy skills and enjoy helping people? Financial advisers work with individuals or companies to help them make good decisions about their financial investments and pensions.
- **Actuary** – Are you good at maths and interested in business and economics? This role is in a range of industries, for example, banking, insurance and investment management. They use statistical modelling to predict the chances of a specific event occurring and the associated financial costs.
- **Financial analyst** – Do you enjoy maths and learning about business? Financial analysts have an excellent understanding of the economy, as well as a related degree such as statistics or economics. They review a company's finances and help them to make good financial decisions.

- **Auditor** – Have you thought about becoming an accountant? Auditors are chartered accountants who specialise in reviewing company accounts to make sure they are compliant with financial laws. They may also advise the company to help them to manage risks and make improvements.
- **Forensic accountant** – Do you have excellent mathematical skills and a strong sense of right and wrong? This job involves investigating financial misconduct. An accountancy qualification is needed as well as attention to detail and excellent communication skills when producing and presenting the findings of your investigative work.

Within the financial sector, artificial intelligence (AI) is affecting low-skilled jobs, such as data entry. The work of more skilled professionals is also changing, as AI can be used to identify patterns and investment opportunities. However, AI cannot replace skilled professionals who can interpret the patterns in the data and use their expertise and communication skills to advise customers to make sound financial decisions.

For more information, visit the Financial Careers website (www.efinancialcareers.co.uk/).

Health

Health is one of the largest sectors in the UK and involves a huge range of jobs. Some are directly involved with patient care and others are focused on research, technical and administrative support roles.

The NHS is the UK's biggest employer and its career website promotes over 350 different health-related careers. Other areas of employment include social care, large companies (see the information box about careers at Specsavers in Chapter 8 as an example), research and development within universities, and SMEs. (See the box in Chapter 3 for information about SMEs.)

These five examples of jobs in the health sector focus on science jobs. There are also many other opportunities to work in the health sector, for example in finance, business support and IT roles.

- **Advanced nurse practitioner (ANP)** – Do you enjoy biology and like the idea of helping people? This is a highly trained role involving diagnosing illness, prescribing medication and helping patients to manage chronic illness. A nursing degree and postgraduate study is required. Both hospitals and general practices employ ANPs.
- **Healthcare assistant (HCA)** – Would you like an entry-level role that can be started after leaving school or college that involves looking after people? As the job involves caring for patients, good communication skills are required. HCAs work in hospitals, care homes and general practice. There is the opportunity to train and gain further qualifications.

4 STEM sectors (helping others)

- **Occupational therapist (OT)** – Are you compassionate and able to work well with others? An OT provides practical support to people to help improve the quality of life and adjust to disability or illness. They work with a range of people including babies, children, the elderly and those with long-term physical or mental health conditions.
- **NHS scientist** – Do you enjoy science or engineering and plan to study for a degree? The NHS science graduate training programme is a three-year scheme that can lead to senior healthcare science jobs in a range of disciplines, including biomedical engineering, genomics, nuclear physics and audiology.
- **Genetic counsellors** – Are you compassionate, meticulous in paying attention to detail and interested in biology? This role involves using medical genetics and counselling skills to support patients, and their families, with a genetic condition diagnosis. A degree in a related subject and a postgraduate qualification, such as a master's degree, is required.

Current developments in healthcare offer exciting career opportunities. New technologies, involved with data handling and medical devices, are advancing medical treatments. The challenges of caring for an ageing population means that problem-solvers and innovators with good scientific, technical and mathematical skills will be needed.

For more information, visit the NHS careers website (www.healthcareers.nhs.uk/) and see the information box about NHS graduate schemes.

MORE INFORMATION ABOUT . . .

NHS graduate training opportunities

The NHS is the largest employer in the UK, with over 1.7 million staff! The organisation has about 350 different career roles including health professionals, scientists and business support. The NHS has a variety of graduate training schemes, including:

- accelerated programmes, such as speech and language therapy, for those with a related degree
- NHS Graduate Management Training Scheme, specialising in areas such as finance
- NHS Scientist Training Programme, with senior roles in a variety of areas including genomics, biomedical engineering and medical physics
- NHS Graduate Digital, Data and Technology, for example, focusing on health informatics.

For more information, visit www.healthcareers.nhs.uk/career-planning/study-and-training/graduate-training-opportunities.

Logistics and transport

Think about how you got your food today, the books that you need for learning and the devices you use. Transport and logistics are involved with getting products efficiently from A to B. It involves planning to safely minimise the time, cost and environmental impact of the transportation of goods around the world.

Five examples of careers in logistics and transport are:

- **Transport/town planner** – Would you like a career involving creativity, good problem-solving skills and the ability to think about the user's experiences? This job involves working on projects to design, improve and implement town developments and transport schemes. A degree in a related subject such as geography, maths or civil engineering is required.
- **Supply chain manager** – Are you logical, able to plan and stick to a deadline? Supply chain managers are responsible for getting the raw materials to the factories and the goods to the retailer's shelves. A good understanding of consumer behaviour and global events are required in order to correctly predict the quantities of goods required.
- **Freight forwarder** – Do you have excellent communication and planning skills? In this job you plan the most cost-efficient ways to move goods (the freight) from A to B using IT and numeracy skills and knowledge of geography. Negotiation skills are also needed to work with clients (who own or are buying the goods) and the rail, road, air and sea transportation companies.
- **Air traffic controller** – Are you able to handle pressure and respond to challenging situations? This role requires good numeracy and technical skills. They are responsible for the safety of aircraft from take-off, during flight and when landing. NATS is the organisation that oversees traineeships in the UK and provides services across all UK airspace.
- **Merchant navy officer** – Would you like a career at sea that does not involve the military? Join the merchant navy to specialise to become an officer within navigation, engineering or technical routes. This apprenticeship involves study at college as well as time at sea. This can lead to a variety of maritime work on ferries, container ships, luxury yachts and cruise ships.

The logistics and transport sector often gets overlooked when people are thinking about future careers. However, this sector is vital to the smooth running of the economy and society. Remember people running out of toilet paper and hand sanitiser in March 2020? The pandemic and conflict in Europe disrupted supply chains, but suppliers and transportation services were able to respond quickly to make sure communities have what they need. From drivers and warehouse

4 STEM sectors (helping others)

operatives to roles that require more training and qualifications, we all need the logistics and transport sector.

For more information, visit:
Logistics UK, https://logistics.org.uk/
United States website Logistics for Kids, www.camcode.com/blog/logistics-guide-for-kids/

Information technology (IT)

The IT sector supports all areas of business and public services. The three ways to view IT in the workplace are:

1. To enable you to do your job, for example, using a computer for communicating or processing information.
2. Using software to help problem-solve, for example, finding patterns in data. Many of the professional careers highlighted throughout the sectors profiled in Chapters 3 and 4 require this level of IT skill.
3. Using specific technology knowledge to create applications and systems. The five professional technical careers listed below require this level of technical expertise. These jobs require technical qualifications from college, university or gained via an apprenticeship.

Five examples of professional technical careers in IT are:

- **Software engineer** – Do you understand why programming languages work and enjoy writing code? This role uses engineering principles (design, develop, test and maintain) and computer programming knowledge to build new software for others to use. They are involved in making products for everything from mobile phones and cars to games and apps.
- **User experience (UX)/User interface (UI) designer** – Do you have excellent problem-solving skills and computing skills? The UX designer uses coding to make a product such as an app or website. They design, test and refine the product to make the user experience as good as possible. The UI designer uses coding to make sure the product looks and sounds good, for example on different screens. These two jobs are also important in the gaming industry.
- **Data engineer** – Do you have good technical skills? Data engineers design, build and test systems that collect, store and manage large amounts of data for the data scientist to use. (Data scientists turn large amounts of raw data into meaningful and useful information. This can then be used by the company to find patterns and improve performance.)

- **Security architect** – Are you interested in cyber security? This role involves designing security systems for a company's IT network. It is important that a company's data is stored safely to prevent cyber-attacks and data leakages. The demand for cyber security jobs is increasing as data has become a valuable commodity.
- **Cloud engineer** – Do you have good programming skills and understand how the internet works? Cloud engineers are responsible for the design, implementation and management of a company's computer services. This includes servers, data storage and networking. The demand for cloud engineers is increasing as more companies move their computer services from an on-premises model to internet-based resources.

If you are interested in understanding how the applications and software packages you use work and like learning coding, the careers above may be ideas for you to consider.

Professional technical IT people may be employed in a specialist company that provides services to other businesses and organisations. If you have excellent IT skills and are interested in a specific sector, for example conservation or health, you may find a meaningful job that requires your skills and reflects your values. (See Chapter 5 for more information about why your values are important in the workplace.)

Read Talal's story as a business intelligence analyst in Chapter 5, and see Catriona's story at the end of this chapter to discover how having a maths degree helped her career as a systems analyst.

For more information, visit the Tech Jobs website (www.technojobs.co.uk/).

The value different industries contribute to the UK economy

The Office for National Statistics (ONS) collates data about the economic output and the total number of jobs for different industries. This information is presented in research-briefing papers. The images below have been taken from the 2022 report. Gross value added (GVA) is a measure of the industry's contribution to the economy.

The ONS is the largest provider of statistical services in the UK. Visit the website for information about working there and career opportunities (www.ons.gov.uk/).

4 STEM sectors (helping others)

Figure 1 The measure of the industry's contribution to the economy. The pale bars are classified as service industries by the government. The real estate measure is misleading as it includes the hypothetical rent owners would pay if they rented their property. Which industries are you interested in? What proportion do they contribute to the economy? Look at figure 2 to find out the proportion of jobs in each industry.

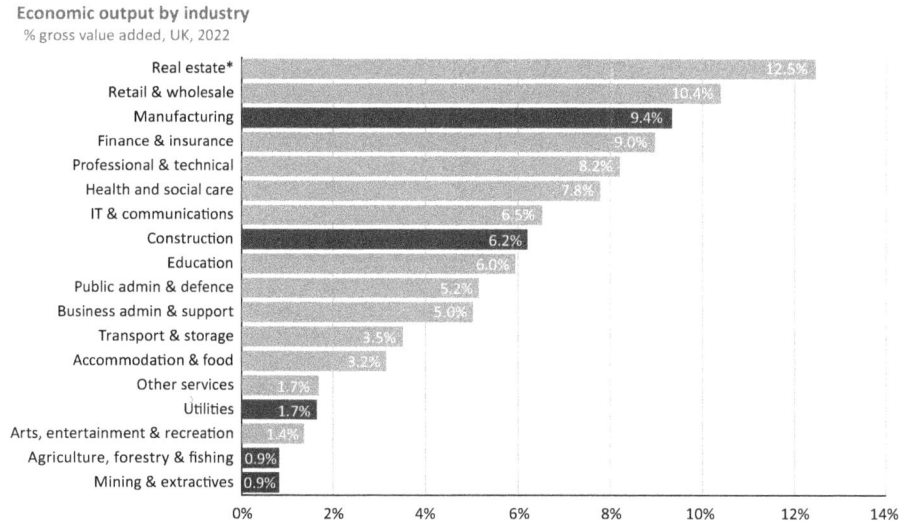

Figure 2 The proportion of jobs in each industry. How does this chart compare with figure 1? Don't let the data put you off exploring careers in an industry. Instead, make sure you carefully research the career pathways and job opportunities so that you are well-prepared and have realistic expectations.

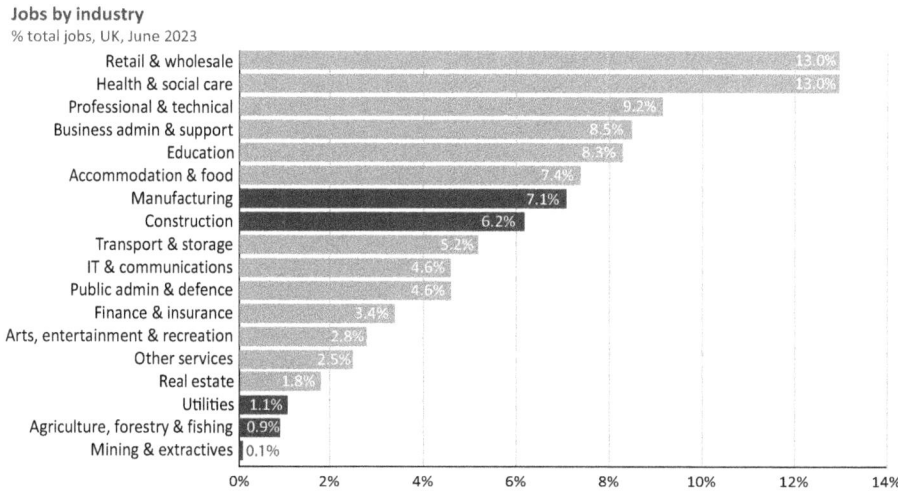

Source: https://commonslibrary.parliament.uk/research-briefings/cbp-8353/#:~:text= Manufacturing%20industries%20contributed%20%C2%A3204%20billion%20in%20GVA %2C%2010%25,sector%E2%80%99s%20output%20was%20%C2%A3124%20billion%2C %206%25%20of%20GVA.

Contains public sector information licensed under the Open Government Licence v3.0.

Fascinating fact

Conservationists across the UK are working with landowners to *rewild* areas of our countryside. Whole ecosystems can be restored so that people and nature can live in balance. The effects can be far-reaching. Reducing grazing from sheep in upland areas and allowing more trees to grow can slow down the movement of water to lowland towns, reducing flooding.

Conclusion

It isn't possible to list *every* STEM-related job in a book of this size. There are 60 jobs highlighted in Chapters 3 and 4, and a further 60 jobs in the A–Z section at the end of this book.

Many websites have detailed information about hundreds of jobs, and a really good one to look at is Prospects Job Profiles (www.prospects.ac.uk/job-profiles). This A–Z collection of profiles are written by professionals, including careers advisers, and contain lots of good advice about entry requirements for the job, the skills needed and sectors, industries and organisations that employ that role. The *UCAS website* also contains information about jobs within different sectors in their 'browse careers' careers section (www.ucas.com/explore/career-list).

 REFLECTION ABOUT CHAPTERS 3 AND 4

Your ideas from reading Chapters 1 and 2 may be useful, but don't worry if you haven't done the reflection pieces. Chapter 3 is about STEM sectors that 'make things', that is products, and Chapter 4 is about STEM sectors that 'help others' that is offer services.

- Which of the *twelve sectors* make products or offer services that you are really interested in?
- Which *companies* make the products or offer the services you are really interested in?

Have a look at one or two of these organisations' websites and try to find answers to the following:

- What is the purpose of the company?
- What career opportunities are there at the company?
- Does the company employ trainees and apprentices? Are there graduate opportunities?
- Can you see any job titles that are of interest to you?

Keep a note of your thoughts as they will be useful in Chapter 9. (What do job adverts tell us?)

Career story: Catriona

Systems Analyst (retired), Standard Life, Edinburgh

Describe your early maths career

When I was at school I had a thing for maths and pretty much found it effortless, much to the annoyance of my friends. I struggled with English and had to retake the Higher exam. I was keen on physics too and thought petroleum engineering at Imperial College sounded exciting, but I got a Grade B when I was predicted an A, so thought I might struggle at uni. Because I found maths so straightforward, I decided to go with that subject for my degree.

I did a maths degree at Herriot Watt University. Half of the people on the course did actuarial maths, but I knew I wasn't into financial wizardry and wanted something different and enjoyed the logic of maths. I never regretted my decision to stick to it and remember doing two pages of integration and the answer was just πi, and I thought it was really funny, all that effort just to get πi!

By the end of my degree, I thought I had pushed my maths as far as I wanted to. Technology as a career was just taking off and I wanted to go in an IT direction. In the career books, I found systems analyst. I didn't know what it was but it seemed vaguely ITish.

What do systems analysts do?

We solve a business problem, for example, a business may need to change its IT. The analyst scopes out what needs to be done and manages the details to solve the problem, limiting the 'scope creep'.

'Scope creep' is when the problem isn't clearly defined and things keep getting added. I worked with business IT systems, but let me explain this with example:

Someone wants a new coffee machine, so you find one for them. Then they complain that it doesn't make frothy milk. But they never asked for that. So you now have to find a coffee machine that also froths milk. You have wasted time and money because of 'scope creep'.

So the best question to ask is, 'What is problem are you trying to solve?'

With the coffee example, this could be 'I want to be able to drink frothy coffee at home'.

It is all about listening and asking questions, solving problems to work out how things can work better that fit within the project budget. This involves a lot of logic to work out what we are able to do.

One of the best courses I did was about influencing skills. This was eye-opening and helped me learn about listening, interpreting and speaking. I was naturally passive, but this helped me to build my self-confidence, communicate better and become more assertive at work, and this was liberating.

I became a senior analyst and at any one time was the lead on one large project or two to three smaller projects. As I was in a senior role, I managed other analysts, helping with their career development.

From work to 'retirement'

I knew I wanted to retire from Standard Life at 55. In my last couple of years, I worked four days a week. I stopped doing people management and focused on being the lead on big projects.

I left in January 2020 with no real plan of what retirement would look like and six weeks later we were in lockdown. I started bookkeeping for my husband's business because his accountant was unable to do this during the pandemic.

Through my husband's business I met the director of a local charity, and I now volunteer with them and helped them set up their bookkeeping systems. This is a good way of using my professional skills to help a charity.

What advice would you give to someone at school?

Find what interests you. If there is a subject that you study and you lose time because you enjoy it so much, they are the ones to go for. This will make it much easier when you study and work.

None of my career would be possible without a maths degree. I haven't done an integration in over 35 years, but maths teaches you to be logical and follow processes.

BE YOUR OWN CAREERS RESEARCHER

Reading Chapters 3 and 4 may have given you some new career ideas or reinforced your interest in possible jobs for the future.

You now need to become your own careers researcher and find out more about possible careers that interest you. This is a bit like being a detective. You need to cross-check facts to make sure your sources of information are reliable.

Talk to people

You can talk to family and friends, careers advisers and subject teachers. Chapter 10 has more information about how these people can help you. By talking to the people who support you, you may be able to make contact with people who actually do the job you are interested in. Ask for help with arranging to make contact and planning what questions you can ask.

Use the internet

- **Have a clearly defined research question.** Don't lose hours on a vague internet search. Give your search structure, for example,
    - What work experience would help me to learn more about working in this sector?
    - What exactly does a [job title] do?
    - What qualifications are needed to become a [job title]?
    - Which companies employ [job title]?
- **Look at company websites.** Have a good search around the website and social media to read about what they do. Do they advertise jobs and can you see jobs you are interested in?
- **Visit websites that represent the sector.** Many sectors and industries in the UK have an organisation that represents them. This can be a very useful source of information for you. Many of these websites are listed in Chapters 1, 3 and 4.
- **Look at college and university websites.** There will be information about the qualifications that you may need for a specific career. What are the prerequisites for applying for that course? (See Chapters 6–8 for more information about qualifications.)

STEM Careers

> **Find out about the job from people who do the job.** Read case studies and job profiles and watch videos of people talking about their job. Find videos on many careers websites including BBC Bitesize Careers (www.bbc.co.uk/bitesize/careers) and Icould (www.icould.com/explore).

5 Useful skills and personal qualities for STEM roles

Introduction

This chapter will look at skills and values in detail.

There will be things that you are good at and others that you are not so good at – these are skills. You can learn new skills and improve the skills you have. So, if you think you are not very good at working with other people, this is a skill that you can choose to develop. You may hear teachers, careers staff and visitors to the classroom talk about skills.

There will be principles that are important to you such as being kind or living sustainably – these are your values. Your values may stay the same throughout your life or some might change as your life circumstances change. So, kindness may be a lifelong value but your financial circumstances may affect your sustainable living decisions at different times in your life.

What does the term 'skills' mean?

Organisations need employees with a range of knowledge and skills who will work well together to achieve the aims of the company. As well as qualifications required, employers will want candidates to have particular skills. Look at any job advert and you will see phrases like 'attention to detail', 'communication skills' and 'work well within a team'.

You'll see and hear lots of vocabulary in the 'skills discussion', such as 'essential skills', 'transferrable skills', 'soft skills', 'employability skills' and 'green skills'. It doesn't really matter what language your school or college uses, as long as you have an understanding of why skills are important, what skills you have and how you could develop your weaker skill areas.

Skills can be broadly grouped into three:

- skills linked to your **knowledge** and how you perform tasks and solve problems

- skills linked to how you manage your emotions and performance at work (**self-awareness**)
- skills linked to how you **work with other people**, both within your organisation and externally.

Here are 12 skills, listed within the three groups above, with a brief description about why they are important to employers. There are also questions and tasks to help you reflect on your strengths and develop your weaker skills. Your skills aren't 'fixed', with a bit of practice, they can be learnt and developed.

Skills linked to knowledge

Problem-solving

This means being able to identify a problem, consider different solutions, try them out and evaluate which is best. This is an important skill in many STEM jobs. The practical work you did in many subjects, such as science, technology, PE and music, may have given you an experience of this. The problem to be solved may not always be a physical one. Finding the best way to create a piece of writing, get to the next level in a game and juggle homelife and studies all involve problem-solving skills.

Someone with a disability can experience the world differently and may find novel solutions to a problem. Employers value staff with good problem-solving skills because they will help the business to work better. For example, this could be finding a better way to manufacture a product. A company may need to improve the costings or timeline for a project. Numerical skills will help with this. Another problem may be poor product sales. Good communication skills will be needed to find a solution for this.

 REFLECTION ON PROBLEM-SOLVING

Are you good at problem-solving? If yes, give an example of when you've successfully used this skill:

- State the problem and how you identified it.
- What possible solutions did you come up with?
- How did you decide which solution was the best?

Do you think problem-solving is a skill you should develop? Practise by identifying a small problem that is of interest to you, for example, how are you going to fund driving lessons? How can you find enough time to study? Think of a few solutions and try them out. How can you tell if the problem is solved? Which solution helped to solve the problem?

Critical thinking

This means not just accepting things but questioning so that you can make an informed decision. Good research skills are needed to be a critical thinker. Let's use the example of having a part-time job when still at school. Some people who support you may think this is a good idea as it will give you extra money and help you to develop your skills. Others may advise that part-time work will negatively affect your studies and possibly your exam grades. A critical thinker will research the topic and examine the evidence from both sides of the argument. They will then develop their own argument and use reasoning to defend this. You probably use critical thinking every day, for example, making the decision about if a message is spam or actually from one of your contacts.

Employers need their workforce to be critical thinkers, and this is linked closely to problem-solving skills. A critical thinker will identify a problem, for example, by thinking 'why do we do this task this way? Is there a better way?' A critical thinker can help to protect the company, for example, by questioning the intent of an external email and not clicking on any links. Critical thinking is so important. It can actually be studied at school, college, university or in the workplace and lead to qualifications.

 REFLECTION ON CRITICAL THINKING

Are you a critical thinker? If yes, give an example of when you've successfully used this skill:

- What was the focus of your critical thinking?
- What seemed the obvious conclusion?
- How did you challenge this?
- What conclusion did your reasoning lead to?

Do you think critical thinking is a skill you should develop? Practise by challenging something that would be easy to accept, for example, an email from an unknown source.

- What evidence is there that the sender is legitimate?
- How can you double-check this?
- What conclusion do you make and why?

Creativity

Creativity is using your imagination to create new ideas and solve problems. There are many subjects you study at school that help you to become more creative. The obvious ones include art, music and drama. But the tasks you are set in modern foreign languages, English, maths and technology will also help you to develop your creativity. Activities you do out of school can also provide opportunities for you to be creative, such as taking part in sports and clubs, your interests and hobbies and your responsibilities around the home.

All industry sectors value creativity. Being creative shows that you are flexible in your thinking and you can generate solutions to problems and come up with new ideas. Many STEM employers encourage creativity as this can help employees to feel motivated and valued in the workplace. All of this can help the organisations to reach their aims, such as increased productivity or engagement with consumers and clients.

Creativity is closely linked to problem-solving and critical thinking. The Indeed job website gives examples of being creative in the workplace (www.indeed.com/career-advice/career-development/creativity-in-the -workplace).

REFLECTION ON CREATIVITY

Are you creative? If yes, give an example of when you've successfully used this skill.

- What was the problem to be solved or the task to be done?
- What new ideas did you have to complete this?

Do you think creativity is a skill you should develop?

Would you like to develop your creative skills to physically make something?

Do you need to become more creative in order to solve a problem, such as finding enough time for studying and interests?

Numeracy

The UK's National Numeracy organisation says that 'numeracy is having the confidence to use basic maths at work and in everyday life'. This means that as well as having basic maths skills, you are also confident when using them.

At school and college you will be encouraged to do your best in maths qualifications. Some people find this easier than others. It is important to know that having good numeracy skills is not the same as gaining a high grade in a maths exam. However, please work hard to get the highest GCSE grade (National 5 in Scotland) that you can! This is important because many courses and jobs have minimum grade requirements.

Employers need their workforce to feel confident with using numbers. National Numeracy website explains what good numeracy skills are (www.nationalnumeracy.org.uk/what-numeracy).

No one wants to pay more than they should when buying something or miss a connection when working out travel time. Feeling confident with numbers may come naturally to you, or this may be a skill that you need to work on. Try not to use negative language such as 'I can't do maths' and challenge others when they dismiss the importance of good number skills. If your numeracy skills need a boost, check out the National Numeracy Challenge online (www.nationalnumeracy.org.uk/what-numeracy/challenge).

REFLECTION ON NUMERACY

Do you have good numeracy skills? If yes, what is the evidence to back up this claim?

- Qualifications and grades?
- Specific example of when good numeracy skills helped you?

Do you think numeracy is a skill you should develop?

- Have a go at the National Numeracy Challenge.
- Practise solving the types of questions you got wrong.
- Try to use your numeracy skills in everyday life, for example, when shopping.

Technical

This means using specific knowledge to do specific tasks. These tasks can be physical, such as changing a fuse in a plug, or digital, such as using a computer software like Excel. There are lots of opportunities within education and life outside of school for you to develop your technical skills. In technology lessons have a go at using the equipment when you have the opportunity, for example, 3D printers. You'll learn how to work safely and become more confident. Practical sessions within science lessons can also involve using equipment and technology, for example, the centrifuge in biology. In subjects such as art, music and drama you may be able to develop your technical skills by using digital workstations.

Different areas of STEM will require different technology skills. By knowing how to use a variety of technical equipment you can show potential employers that you can adapt and learn new skills. If you know what area of STEM you would you like to work in, try to develop some of the technical skills linked to the sector. So, if you would like to be an electrical engineer, have a go at using a soldering iron to connect components, build a circuit and make a working device. If you would like to work in healthcare, learn some basic first-aid skills and how to take someone's temperature and blood pressure. In Chapter 2, read about how Molly's interests helped her develop technical skills that she used in her podcasting career.

Just about every job will require some level of digital skills, and the box below has more information about 'digital literacy'. The world of work is changing so rapidly, and many employers worry that their staff won't keep up with the changes. Know how to communicate effectively online using a variety of platforms. Be prepared to learn how to use new software. Employees with good technical skills are confident to use what they have learnt from previous experience to trial and adapt to new situations. Don't be afraid of asking for help if you're unsure of how to use an application or device.

MORE INFORMATION ABOUT . . .

Digital literacy

As with all skills, your digital literacy can be developed. Learning how to check the reliability and evaluate the information gained from websites and social media is an important skill. Learning how to use social media responsibly and understanding the consequences of ill-judged posts are important, too. When you are applying for a job, employers and recruiters may look at your social media posts.

In the workplace, you may be required to develop your digital literacy skills further. Communicating online with colleagues and hybrid working (working at home and in the office) have developed immensely because of the pandemic. You may be asked to use a range of hardware and software and be able to learn how to use the updates and upgrades when they are available.

More and more professional interactions occur using technology, and you will be expected to know how to network (build professional relationships) and use online platforms appropriately.

Finally, using the digital world safely is vital, keeping you and your colleagues safe. You should know how to keep personal data secure.

> **REFLECTION ON TECHNICAL AND DIGITAL SKILLS**
>
> Do you have good technical and/or digital skills? If yes, what is the evidence to back up this claim?
>
> - Qualifications and grades?
> - Specific example of when good technical and/or digital skills helped you?
>
> What specific technical skills you need to develop to help you pursue your dream career?
>
> - How can you do this?
> - What digital skills do you need to develop?
> - How can you achieve this?

Skills linked to self-awareness

Reflecting on progress

Do you ever stop to think about how well you are doing at a certain task?

An example could be subject coursework or an assessment piece. Perhaps things are going well. Stop to think why this is so? Are you particularly good at something, for example, planning how to use your time effectively. Maybe things are not going well with the task. Could things be going better? What leads you to think this?

When you ask yourself these types of questions, you are reflecting on your progress. This is a very useful skill because too often people stay on a course of action, even if it won't lead to the desired outcomes. This can lead to failing the task.

In the world of work, you need to be able to 'check in with yourself' and reflect on progress. Are you doing what was asked of you? Is it going well, or could you be doing things better? How do you know this?

Goal setting

Do you sometimes do something without knowing why? How do you keep yourself motivated to work on the task? How would you know when you've finished the task?

Goal setting can help you to stay motivated to complete a task. You define the endpoint of the task and the reason for doing it. This could be as simple as completing a 'to do' list so you have more free time. Or a more complex goal such as gaining a particular grade in a subject so you have the qualification you need for your next career step. Part of goal setting is to have an action plan with the steps you need to follow in order to reach your goal. Having an action plan keeps you focused on completing the task and not waste time on activities that stop you from reaching your goal.

Employers want their workforce to stay motivated as this helps the organisation to achieve its aims. Having goals will help you to stay motivated. These could be daily goals linked to tasks. They may be longer-term goals linked to your professional development or specific projects.

Chapter 10 has more information about setting goals and planning for your STEM career.

Adaptability

This means being able to adapt to new situations and conditions. How well do you adjust to changing circumstances in school or homelife?

Does it take you time to feel comfortable when conditions change? This is OK, as some people can take a while to adjust to change. To improve your adaptability skills, try to become more flexible and acknowledge that change happens. Things can't always stay the same. Be curious about the changes and find out why they are occurring. Try to keep in control of the factors that you can control (e.g. your time management) and be confident that you can respond and thrive in this new situation.

The world of work is changing rapidly as new technology is developed and external factors, such as resource supplies, alter. This can affect business and employers need their workers to be willing to embrace change and learn new methods. Of course, remember that critical thinking is also important. It is OK to ask why the change is occurring as long as you have an open mind and consider how the change can be beneficial to the company and its employees.

Autonomy

Autonomy is making decisions for yourself and being able to work independently. When you were a baby, you had very little autonomy as your carers will have made decisions about your food, clothes and activities. As we get older, we are able to make more decisions for ourselves. Choosing the clothes to wear, what to eat for lunch and how to spend your free time are all important steps in learning to be autonomous.

Here are two ways to think about being autonomous:

- Doing a task – you can complete the task without continually needing to be told what to do.
- Making a good decision – you base this on your research and reasoning and not on what someone else tells you to do.

The people who support you (family, carers, friends, teachers and managers) can help you to learn to be more autonomous. Find opportunities to make your own decisions. You'll learn what worked well and why other things did not. This will help to increase your confidence and you'll trust that you can make good decisions that are true to your values. (See later in this chapter for more information about values.)

 HERO IN STEM: SIR JAMES DYSON

'Enjoy failure and learn from it. You never learn from success.' This is a famous quote from James Dyson.

Sir James Dyson is an inventor and engineer, who started out studying architectural design at the Royal Academy of Arts. His early work involved designing the Sea Truck, and it was his innovative reinvention of the wheelbarrow, the Ballbarrow, that led him to become an entrepreneur.

Resilience and perseverance are qualities of James Dyson. He had the idea of creating a vacuum cleaner that didn't need a disposable bag to hold the dust and dirt. (Ask someone older about the old-fashioned vacuum cleaners!) It took four years and thousands of prototype models to create the first bagless vacuum cleaner that uses cyclonic technology. However, manufacturers rejected his designs as the market for vacuum cleaner replacement bags was worth hundreds of millions of pounds!

Not one to give up, Dyson started his own manufacturing company, and the rest, as they say, is history. You are likely to see Dyson products in your home and in other locations (he invented the Dyson Airblade hand dryer in 2006). Not all his projects work (e.g. his electric car ideas) but Dyson believes that we can learn from failure.

Sir James Dyson is a philanthropist and supports scientists and engineers with their research and development. In response to the shortage of engineers in UK, in 2017 he opened the Dyson Institute of Engineering and Technology, a new type of higher education establishment.

Autonomy in the workplace is important. Your manager will want you to be able to work by yourself and know when to ask for help. This won't happen straight away, and you will probably have training and perhaps a 'buddy' or mentor.

 REFLECTION ON SELF-AWARENESS

Preparing for an exam will be used to help illustrate how you can develop the skills of reflecting, goal setting, being adaptable and able to work autonomously.

- Decide which subject to focus on.

Reflecting

- How have you been doing so far?
- What is going well/not so well?
- How can you check your progress?

Goal setting

- What grade would you like to get in the final exam?
- Why do you want to achieve this grade?
- How long have you got until the exam?
- What steps can help you get to your goal (mock/preliminary exam prep, revision sessions, planning set studying time)?

Adaptability

- Is anything getting in the way of you learning this subject?
- Is anything getting in the way of studying for the exam?
- Why are these things getting in the way, and what can you do about them?
- Do you need to adjust your goals? Have you made a reasoned decision?

Autonomy

- Who can support you to achieve your goal and what can they do to help?
- How much do you need their support and what can you do by yourself?
- How will you know that you can manage by yourself? (This feeds back to **Reflecting**.)

5 Skills and qualities for STEM roles

Skills linked to working with others

Communication

This involves writing, speaking, listening and non-verbal communication skills. In all of your subjects, your teachers will be giving you opportunities to develop these skills. The clubs, hobbies and interests you have may also provide opportunities to develop these skills.

In most jobs, you will be expected to write clear and concise communications, for example, in emails or project reports. Answering a question in a lesson may be difficult for you, but in the workplace, your considered contributions in a meeting will be valued. Listening carefully to others is important. This is one way you'll get instant feedback in the workplace. It is also how you'll find out about your co-workers' ideas. Body language can also give you clues to how someone is feeling. People who feel excluded or not listened to may sit in a certain position. *Your* body language is also important as this will give signals to those around you about how you are feeling and whether you are paying attention when someone is speaking.

 REFLECTION ON COMMUNICATION SKILLS

Ask those who know you well to give you feedback on your communication skills (writing, speaking, listening and non-verbal).

Remember, these skills can all be learnt and developed. So, if you feel a little shy talking in front of a group of people, try to take opportunities to practise in situations you feel comfortable, such as in a favourite lesson or club.

Teamwork and leadership

This is more than being part of a sports team or being the boss! Even if you are the type of person who works best on their own, it is important to learn how to work well with others.

Throughout school, you have opportunities to work as part of a team, for example during practical work in science. You learn to listen to others, take on board their feelings, contribute your ideas and help the team to achieve the task. Leadership can be formal, such as being a designated team captain. In this role, you learn to manage a project by breaking it into tasks, delegating the workload and encouraging everyone to contribute. Leadership can also be informal, for example, encouraging a friend to finish their homework. Whether you enjoy taking part in group activities or not, it's all practice for the world of work.

No job is ever done in complete isolation. Employers want employees who can cooperate with others to successfully complete the task. Staff need to be able to work with a variety of people including colleagues, customers, clients and people from organisations external to the company. Not everyone can be the boss, but we can all demonstrate good leadership skills. An example of informal leadership in the workplace is having good work behaviours such as good time keeping and being organised. Another example is being open to new ideas and embracing change in a positive way. In your early career, you may be encouraged to develop your teamwork and leadership skills by being involved in small projects.

 REFLECTION ON TEAMWORK AND LEADERSHIP SKILLS

The *Skills Builder Framework* gives guidance to help you develop the essential skills for work. The steps for developing **teamworking** skills include:

- working well with others
- contributing to a group
- improving a team
- influencing a team.

For ideas about how you can develop your teamworking skills, visit the Skills Builder website, which also includes steps to help you build many of the skills discussed in this chapter (www.skillsbuilder.org/framework).

Negotiation

This is an important skill that often gets forgotten. Can you remember asking for pocket money or a later bedtime when you were younger? Did you get what you wanted, and why was this? These are examples of times when you have been developing your negotiation skills.

A negotiation is a discussion between two or more individuals/groups about a particular thing that leads to an agreement that is acceptable to everyone involved. In order to achieve this, both sides must prepare, present their case and listen to any opposing views. If common ground can be found, both parties may be happy to compromise so that a solution can be found.

Let's take the example of asking for more pocket money. You may have reasons for needing the money, but your family are unlikely to have lots of extra cash to give you. But if you can offer to do an extra chore around the house and your busy parents/carers see the benefit in having one less task to do, they may be willing to increase your allowance.

It is similar in the workplace. You may need to negotiate your working terms and conditions. Or your day-to-day work may involve negotiating with other colleagues and external organisations so that a project is delivered on time and to budget. The skills required are similar to those you used when you arranged extra pocket money or a later bedtime!

 REFLECTION ON NEGOTIATION SKILLS

Think about a time when you tried to persuade someone (or got your own way!).

1. Did you get the outcome you wanted?
2. Was the other person happy with the outcome?

- If the answer to both of these questions is 'yes', how did the discussion lead to you and the other person finding the outcome acceptable?
- If the answer to either of questions 1 and 2 is 'no', how could you have done this differently so that both parties were in agreement with the outcome?

Read the career stories throughout the book to find out how people's skills are important in their day-to-day work. Lauren discusses using her negotiation skills when she wanted her employers to support her to study for a degree, read her career story in Chapter 1. Talal offers some great advice about skills in his career story at the end of this chapter.

STEM Careers

Bringing skills together

Think of your skills as your bag of tools.

As with all tools, you need to look after them to keep them in good working order. You also need to practise, so that you can use them effectively.

Skills are all connected. Developing in one area will help you in another. As you become more self-aware, you can develop your knowledge skills and think about how you work with others.

Values

The values and principles that are important in your day-to-day life will probably be important to you at work too, so we will call them 'work values'. Here is a list of 20 work values (note: there are many others):

1. achievement
2. helping others
3. adventure
4. routine
5. money
6. risk
7. excitement
8. belonging
9. challenge
10. competition
11. contact with people
12. creativity
13. flexibility
14. decision-making rights
15. independence
16. social justice
17. learning opportunities
18. predictability
19. security
20. variety

You are more likely to be happy at work if your values are similar to those of the company you work for. If you are content and happy, you are more likely to thrive and be successful at work.

Your values can change over time. As you grow older and your lifestyle and responsibilities change, so can your values. This is worth remembering, as an exciting job with lots of risk and travel opportunities might appeal to you now but may not be important to you if you have caring responsibilities. In Chapter 9, Doyinsola discusses how his values are important in his work life.

REFLECTION ON WORK VALUES

Look at the list of 20 values. Are there any other values that are important to you that have been missed out? If so, add them to your list.

Rank the work values, including any extras you added, in order of importance to you, with number 1 being the most important. The best way to do this is to copy out the work values onto separate pieces of paper, cut them up and move them around to get the order you feel most comfortable with.

In Chapters 3 and 4, you thought about different STEM sectors and companies that operate within sectors you are interested in. How do you find out if their values align with your own?

- Choose one company to investigate and find their website.
- Look around the website for the terms: 'values', 'aims', 'vision' and 'mission'.
- Do you see vocabulary similar to your top ten values? If so, it is likely that the company's values are similar to your work values.
- Next, find a job advert for a role that interests you.
- Can you find any of your top ten values mentioned in the job description?

When you are applying for a job, remember to research how the company's values compare to your own. Chapter 9 has more information about learning from job adverts.

 Fascinating fact

Nuclear energy has been used to generate electricity since the 1950s. A new generation of reactors, small modular reactors (SMRs), is being developed, involving collaboration between scientists, engineers and mathematicians within industry and academia. SMRs will generate electricity, green hydrogen and sustainable aviation fuel, helping the UK towards net zero energy production.

Conclusion

This chapter focused on the skills that are useful in the workplace and why it is important to consider our own work values when researching a possible future employer.

In different organisations the language used to describe a skill can vary, but the important thing to remember is a skill can be in one of three groups:

- knowledge – what you know and can do
- self-awareness – how you think about and manage your behaviour or
- working with others.

Employers value these skills in their workforce, and it is important that you know your strengths, weaknesses and how to develop specific skills. Reflecting on your work values is also crucial. You are more likely to be happy at work if your employers have similar values to yourself. Remember, your work values can change as you get older, and this may be the reason to look for a different job or career in the future.

> **COLLECTIVE REFLECTIONS ABOUT CHAPTER 5**
>
> *Skills*
>
> - Ask your trusted family and friends which skills are your strengths and which need developing.
> - For the skills that need developing, use the guidance in the associated reflection boxes within this chapter.
>
> There are many websites and resources that can support skill development. One that is used by many schools and colleges and endorsed by many organisations is the Skills Builder Partnership. Explore their website and you will find lots of material to help you (www.skillsbuilder.org/framework).
>
> **Work values**
>
> Not everyone is happy in their job, and it may not be easy to discuss work values with your family and friends who are at work. However, if you can find someone to talk about this, you may learn from them about what does (or doesn't) lead to being content at work and doing a job that gives satisfaction.

Career story: Talal

Business Intelligence Analyst, Direct Line Group (DLG)

What is your job?

I am basically a data analyst. I report on metrics, such as key performance indicators (KPIs) for the company on a monthly basis. I also use the data to answer business critical questions for managers in DLG. I use Python and Power BI (a Microsoft visualisation tool) to get deep into the data and check we are receiving the right data from the garage repairers we work with. This all helps our company to work better. Business analytics is quite new so I've had the opportunity to upskill and do a level 4 apprenticeship while doing the job at the same time.

Did you imagine you would be doing this job when you were younger?

No! I come from an Asian family and my dad said I had to become a doctor or maybe a pharmacist. But I didn't get the A level grades needed in the subjects I studied (chemistry, biology, physics and psychology). This gave me some freedom to look for other courses I could apply for with the grades I had. The Foundation degree in engineering looked good because I liked maths. This lasted for a year and gave me a platform to choose from a range of engineering degrees, and I went on to do mechanical engineering. Many of my friends went on to do a master's degree, and I might want to do in the future. After my Undergraduate degree, when I was 20, I got onto the DLG Technical Graduate scheme.

How did you get the Technical Graduate job at DLG?

It was important for me to find a graduate job as I was married and needed to earn a good salary. I applied for lots of grad schemes that needed maths or engineering skills. By the time I had the DLG interview I was not so nervous as I'd had a few interviews so I decided to just 'be myself'. We had assessment group tasks to do and I helped our group by taking the lead and trying not to be self-conscious. I'm good with people and I think I got the job because I let myself shine though.

How did Covid affect things?

In 2020 I was part-way through the graduate scheme. We were rotating through different teams in the company, such as Bodyshop repairs. I had to start working from home and being physically in a department had to change. We graduates worked together and helped each other, and my manager was great. I applied for a job within his team after the scheme and that's where I am now.

What advice would you give to people thinking about their career?

Don't worry about what skills someone else has. Try not to compare yourself with others as this doesn't empower you. Focus on the skills you have and how you can develop them. Keep pushing forward and don't get too comfortable. Take the opportunity to upskill and learn new things. Careers are not linear anymore and you'll probably change jobs and companies. You can use your social media skills on professional networking sites such as LinkedIn to share interesting content related to work and market yourself.

6 STEM qualifications (making choices)

Introduction

The next three chapters are about the options you have during school into further education and employment. This chapter will focus on making decisions. Chapter 7 will focus on the choices you can make that mainly involve studying. Chapter 8 will focus on the choices you can make that mainly involve paid employment (working). It is important to clearly state that there is no one pathway. You will hopefully have a long career, and it will most likely involve periods of retraining and professional development.

Some people follow traditional routes from school into work or into university and then work. Others may move back and forth between studying and working, and some feel lost and unsure about what they can do. It is hoped that by reading this book, you will have fewer times when you feel you don't have a plan.

You may not know what you want to do with your career, and that's OK. But you should always have a plan. A plan may be short term, for example, doing a bit of research about a career or getting a work experience placement. Or a plan may last for a longer period, for example working hard to complete a qualification or learning a new skill such as coding. Chapter 10 will discuss in more detail how you can make plans.

It is important to make it clear that only *you* can make the choice about what to do next. Friends, family, school and other influences can guide you. But it is up to you to decide. So, how do you make a good decision?

Making good career decisions

To make a good decision, you need to have as much information as possible. You can then weigh up the positives and negatives about each option. You've been making decisions since you were a toddler: what to eat, what to wear and what activity to do. Sometimes your decisions can be driven by your emotions. It is important that you understand how your emotions affect your decision-making.

Do you want instant gratification or are you prepared to wait for something?

Do you like to please those closest to you or are you fiercely independent?

Understanding your 'default' emotions when making decisions can help you to pause and weigh things up before choosing what step to take next in your career. This should help your decision lead to a positive outcome that you are satisfied with.

Career steps

You've already started your career! Remember the definition introduced in Chapter 1? Our career is our pathway through life, learning and work. Steps you have already taken in your career may include:

- subject choices for GCSEs or National qualifications (Scotland)
- hobbies, sports and clubs you take part in
- any part-time paid work
- any volunteering you do.

Responsibilities you have at home, for example caring for a family member, are also part of your career. You may not have had much choice, but these experiences are adding to your skills and understanding of the world.

Your next career steps could be into education, employment or maybe both, but what you can do may depend on your age and where in the UK you live. At the end of this chapter, you can read Ruaridh's story about how activities he took part in while at school helped him choose a career pathway and how he compared Scottish and English university courses before deciding on the best option for him.

School-leaving age

In England, you can leave school after Year 11, as long as you have turned 16 before the end of the summer holidays. However, you must then stay in either full-time education, work with training (such as an apprenticeship) or part-time education/training with 20 hours or more each week in volunteering or paid work.

In Scotland, Wales and Northern Ireland you can leave school when you are 16, but the exact date you can leave depends on when you turn

16 and which home nation you live in. There is no legal requirement to do any more training or education after this age, although you will be strongly encouraged to do so.

Study, work or both?

You will hear and see lots of information about academic or vocational routes. Academic is defined as education and scholarship. Vocational is defined as work and occupation. This book does not want to separate the two.

- You can go to university and do a degree while having unrelated paid part-time work. Read Molly's story in Chapter 2 about how part-time work in education gave her skills she uses now in media production.
- You can have a paid full-time job and take unrelated courses in your evenings or weekends to gain qualifications.
- There are college and university qualifications that have formal work placements or extended periods of paid work.
- Some jobs involve training and working towards related professional qualifications.

These can be very muddy waters. Try not to get sucked into thinking you have to follow an academic or vocational pathway. One thing is certain, with the development of new technologies and changing world of work, whatever the job, we will all be required to learn new things during our working lives.

Factors that may influence your decision to stay in education or go into employment can include friends are taking that study route, a family member followed that career route, it's your favourite subject, it does/ doesn't involve travelling and the financial implications.

Some people may not be ready to settle into long-term studying or a job, and they may want to take time out to focus on different experiences before committing to anything longer term. A 'gap year' may be just the thing.

Taking a gap year

A 'gap year' is not just for university students, and it may not last for a whole year. What is important is that the time is used constructively. You may choose to try a range of things that give you new experiences and allow you to develop skills you may want to strengthen. Examples of activities that can be done during a gap period include:

- **Travel**: Planning few weeks in a different country will help to develop your independence.
- **Paid work:** Jobs, for example in hospitality or retail, are a great way to develop your confidence and understanding of what the workplace is like. You can also save money to help with further study.
- **Volunteering:** Helping a local organisation once a week and giving your time up to work at a charity event or festival are great ways to learn about commitment and teamwork.
- **Learning a new hobby:** This is a great way to improve your resilience and self-belief – for example, learning to sew, playing a musical instrument or carving wood.

If you are considering a gap year, it is best to have a plan so that you have clear goals and an endpoint. There must be a purpose to this time, such as saving money, developing skills or becoming more independent. When you feel ready to apply for a STEM course or a job, you will have lots to write about in your personal statement and talk about during an interview.

Unplanned events

You may have a plan for a dream job, but the world is unpredictable and 'life happens'! Unexpected things can affect your life and you may have to change your plans. You can develop behaviours to help protect yourself from the fallout after an unplanned event. Try to approach life and learning with these four behaviours:

- Be **curious**: don't dismiss a new idea or assume you already know everything – ask questions, try new things and meet new people.
- Be **flexible**: you may need to change your plans to get to your desired goal (e.g. a dream job). Or you may need to find a new goal. Try to adapt and see the 'bigger picture'.
- Be **persistent**: making mistakes and failing can be a good way to learn about yourself.
- Be **optimistic**: be positive and hopeful for the future. That revised plan you had to make could lead you to something new. Unplanned events could lead to new opportunities.

As well as helping you to cope with random events, having these behaviours may help you to see new opportunities and increase the chance of lucky events happening to you! Read Josh's story in Chapter 3 to find out how the pandemic led to a change in his career pathway.

Changing pathways

Learning is lifelong and setting out on one pathway does not exclude experiencing another. For example, following a STEM vocational qualification at college can lead to degree study at university. Studying A levels and Highers can lead to apprenticeship. Some people completely change their career direction and you can read about Bonnie's experiences of switching from business studies to healthcare in Chapter 8.

Comparing qualifications

It is important that you know the different qualifications you can take in each stage of your education and in employment. The qualifications differ slightly across the four home nations. Have a look at the tables on the following pages to see the different qualification and levels in England, Wales and Northern Ireland and in Scotland. Education, recruiters and employers may use the levels alongside the qualification name, so it is worth understanding what the levels mean.

Chapters 7 and 8 will give an overview of each qualification. People in your school and college can also help you to find out more. You can talk to careers advisers, teachers and support staff. Also look up the qualifications on the internet by going to college and university websites. The UCAS website is also a good source of information about different qualifications (www.ucas.com/).

 Fascinating fact

To help reduce global emissions, new ways of heating homes and offices are being developed. Even cold air has some energy in it. This energy can be extracted by air-source heat pumps and used to increase the temperature of a fluid in the heat pump system. This heat is then transferred into the home. Skilled engineers will be in high demand to install and maintain the pumps.

TABLE 1: England, Wales and Northern Ireland (NI)

Level	Examples of qualifications, including vocational
Entry	Entry-level certificate, Essential skills (NI), Functional skills (England)
1	GCSE – Grades 3, 2, 1 (or Grades D, E, F, G) Welsh Baccalaureate Foundation Music Grades 1, 2 and 3 Level 1 National Vocational Qualification (NVQ)
2	GCSE – Grades 9, 8, 7, 6, 5, 4 (or Grades A*, A, B, C) BTEC Welsh Baccalaureate National Music Grades 4 and 5 Level 2 NVQ Intermediate Apprenticeship
3	A level, T level, AS Extended Project Qualifications (EPQ) BTEC, Alternative Academic Qualifications (from Sept 2025) Welsh Baccalaureate Advanced Music Grades 6, 7 and 8 Level 3 NVQ Advanced Apprenticeship
4	Higher Technical Qualification (HTQ) (England), including higher national certificates (HNCs) Level 4 NVQ Higher Apprenticeship
5	HTQ (England), including Higher National Diplomas (HNDs) and Foundation degrees Level 5 NVQ
6	Degree with Honours (e.g. Bachelor of Arts (BA) hons or Bachelor of Science (BSc) hons) Ordinary degree without Honours Professional Certificate in Education (Wales) Level 6 NVQ Degree Apprenticeship
7	Postgraduate Certificate (e.g. in education (PGCE)) Master's degree (e.g. Master of Arts (MA), Master of Science (MSc)) Integrated master's degree, for example Master of Engineering (MEng) (see Chapter 8) Level 7 NVQ
8	Doctorate (e.g. Doctor of Philosophy (PhD or DPhil))

TABLE 2: Scotland (Scottish Credit and Qualifications Framework, SCQF)

SCQF level	Examples of qualifications, including vocational
1	National 1
2	National 2
3	National 3
	Skills for Work National 3
4	National 4
	Skills for Work National 4
5	National 5
	Skills for Work National 5
	Modern Apprenticeship
6	Higher
	Skills for Work Higher
	Modern Apprenticeship,
	Foundation Apprenticeship
7	Advanced Higher
	Scottish Baccalaureate
	Higher National Certificate (HNC)
	Modern Apprenticeship
8	Higher National Diploma (HND)
	Technical Apprenticeship
	Higher Apprenticeship
9	Ordinary degree
	Graduate Apprenticeship
	Technical Apprenticeship
10	Bachelor's degree with Honours
	Graduate Apprenticeship
	Professional Apprenticeship
11	Master's degree
	Integrated master's degree
	Postgraduate Certificate
	Graduate Apprenticeship
	Professional Apprenticeship
12	Doctorate, Professional Apprenticeship

Source: Tables 1, 2 and 3 adapted from www.qaa.ac.uk/docs/qaa/quality-code/qualifications-can-cross-boundaries.pdf?sfvrsn=a852f981_16

TABLE 3: Approximate comparison of Scotland's levels with the other home nations

Scotland level	England, Wales and NI level
1	Entry level
2	
3	
4	1
5	2
6	3
7	4
8	
9	5
10	6
11	7
12	8

Conclusion

It is OK not to know what career you want. (Analysis of career paths shows that few people really know what they want to do for a job until they are older and have had more exposure to the world of work.) But it's not OK to do nothing about this!

Research the options available to you and weigh up the positives and negatives of each option. Talk to qualified people, such as careers advisers, as well as family and friends. This should help you to make informed decisions that you are satisfied with. Read how Ethan's careers adviser helped him to find out more about degree apprenticeships in Chapter 3.

Remember that although planning and goal setting are important, pathways are not set in stone. Vocational study and work can lead to academic study at university. It is important to be able to review and adjust your plans when necessary, especially if random events affect your progress. Just because you study a particular subject at university doesn't mean you have to do that for a career. Our university system and jobs market are good at letting you take other paths if you decide that the subject choice is not for you in the long term. Read Doyinsola's story in Chapter 9 about how a degree in engineering has led to other technical roles in his company.

6 STEM qualifications (making choices)

 REFLECTION ABOUT CHAPTER 6

How do you make decisions? Look back at the list of **career steps** you may have taken?

- Why did you choose to do this subject/activity?
- What people may have influenced your decision? Why do you think this occurs?
- Is this a positive or negative influence? Why do you think this?
- How do you feel about the decision to do that subject/activity?
- Would you make that same decision again? Why?

Academic or vocational step next?

What choices *could* you make for your next transition, for example, when you finish GCSEs/Nat 5s (Scotland) or sixth-form studies?

- Make a list of ALL of the options (e.g. work, college, university)
- Add the positive and negative reasons for each option

These reflections will be useful for Chapters 7 and 8.

Career story: Ruaridh

Environment Forester, Forestry and Land Scotland (FLS)

Describe your job

I am an environment forester, which means I work in a team that provides advice on protected species and habitats to our forest planners and delivery teams. This helps them to fulfil the legal requirements with regard to species such as badgers and squirrels and that we are looking after important habitats like peatlands and sand dunes. I also look at various forests and design environmental projects to improve the biodiversity. As part of this, I line-manage a field ecologist who does the survey work.

Every day I am thinking about a particular ecosystem, what we need to do for us and how we can manage this. This could be for timber production, planting a native woodland to improve the biodiversity or improving the chemistry of a water course. I use technology pretty much every day, mainly geographic information systems (GIS). All our land is mapped and I use my phone to make notes about what is happening at an exact location and then manipulate the data back in the office. We also use drones as these can be a really useful tool to observe land quickly and in detail, for example, we can see the damage done after storms.

What influenced you to have an environmental career?

I grew up in a rural village in Scotland, and the Woodland Trust owns lots of land in the area. I was curious about what was going on and how this impacted the land and the community. So I did work experience with them when I was in secondary school. I spent a week with the estate ranger and I learnt about her job. I did some practical work that a ranger does, like litter picking in the car park and making posters for events coming up.

During my final year of sixth form, I had space on my timetable and so spent half a day each week with the Woodland Trust, doing jobs like checking the footpaths and finding invasive species like rhododendron so they could plan their removal. This helped me get to grips with how they manage an estate.

How did you decide on your degree choice?

I really enjoyed the work experience with the Woodland Trust and looked on UCAS to see what courses were available and found the Countryside Management degree at Harper Adams University in Shropshire. When I wrote the personal statement for the UCAS application, I could apply what I'd learnt at the Woodland Trust. As well as studying the required subjects, I had learnt skills about countryside and visitor management that would be useful for the course.

In Scotland, we don't pay tuition fees. But at the time I couldn't find a degree at a Scottish university that interested me as much as I really liked the course and university, so I had to have a loan for the fees. I had a part-time job at Tesco that helped me to run my car and pay the insurance.

Harper Adams is a land-based university and most of the degrees they do are linked to agriculture, the environment, animal science or technology linked to farming. The farm at the university campus was a big draw. We could take what we learnt in the classroom and see it in the field in real life.

This was a sandwich degree and included a placement year in industry. I worked for the Field Studies Council as an education assistant. As this was a placement job, it was excluded from the minimum wage. The pay wasn't great, and I kept the job at Tesco to supplement my income. This part-time work gave me lots of other opportunities to develop skills, such as working with other people and dealing with the public. I was also made a supervisor and learnt how to supervise staff, which is useful for the line manger role I have now.

One exciting career moment . . .

One afternoon I sat for half an hour and watched some otters playing on a river bank. There aren't many jobs when you get paid to spend time doing this.

7 Study (with work experience)

Introduction

As discussed in Chapter 6, there is no one pathway to choose. You can make academic choices and vocational choices throughout your career. To help you find out a little more about your options, this chapter will consider academic choices at the different transition stages of education. Chapter 8 will focus on vocational and work-related choices.

What are transition stages?

Moving from one stage of your life to the next is called a transition stage. Below is a list of some transition stages many of us go through, although not everyone does each at the same age.

- going to primary school
- moving up to secondary school
- going to college
- going to university
- going into full-time employment
- changing jobs
- retiring.

Not all transitions are to do with education. Here are other stages you may go through in your life:

- leaving home
- moving in with a partner
- moving house
- having a family.

Transitioning can be easy for some people and harder for others. Doing your research and weighing up the options can help you to prepare and make good decisions about the transition. Leaving school and going to college, university or work will probably be some of the biggest choices you have to make on your own.

In her career story in Chapter 4, Catriona discusses transitioning from full-time work to 'retirement' and volunteering her business knowledge.

Why stay in education?

There are things to consider when making academic choices:

- What qualifications you want to do and why?
- Do you have the entry requirements for these qualifications?
- Do these qualifications help you with your plans? (See Chapter 10 for more information about making plans.)
- Where is the location of the school/college/university you could attend?
- How will you get there?
- Where will you live?
- What are the costs associated with travel and accommodation?
- How will you fund this?

People have different reasons for staying in education. It could be because you need a specific qualification in order to progress onto a desired career pathway. For others, going away to university can be a stepping stone to independence. Some people may not feel ready to leave home and choose a college or university that they can travel to from their family house. Wanting to meet new people and see new places can also be a deciding factor. You might simply love studying, or a subject, and you are keen to learn more. The desire, or need, to earn money may motivate some young people into looking for employment after finishing school.

The route you decide to take isn't set in stone. If you go straight into work and training, you can come back to full-time or part-time education at any age. (I did my master's degree at the age of 48!)

Types of qualifications

There isn't the space in this book to go into every type of qualification in detail, but the main choices at school, college and university will be discussed. Look at the tables (either for Scotland or the rest of the UK) in Chapter 6 to see the level that each qualification is at.

GCSEs (England, Wales and Northern Ireland), National 5s (Scotland)

You'll get the opportunity to choose some of your GCSEs and Nationals (Scotland). GCSEs are usually taken at the end of the year you turn 16.

In Scotland, you may start taking Nat 5s when you are 15. Choose a range of subjects that you enjoy, may want to continue to study and allow you to develop skills such as creativity and problem-solving. To study STEM-related subjects in sixth form and college, you may need to take these options at GCSE/Nationals. Talk to your subject teachers for more information.

Do your best to work hard and achieve the highest grades you can in these qualifications. Maths and English are important requirements for many jobs and other courses, so aim for at least Grade 4 or above in GCSEs and to at least pass Nat 4 if you live in Scotland. If you are struggling with maths, look back at Chapter 2 for advice. The more you can see the value in maths, the more motivated you will be, increasing your chances of getting a better grade.

AS and A levels (England, Wales and Northern Ireland), Highers and Advanced Highers (Scotland)

You can study AS and A levels after you have completed your GCSEs (England, Wales and Northern Ireland). In Scotland, Highers and Advanced Highers are studied after gaining National 5s. In all parts of the UK you can study these qualifications in school or college, but you will need to check that the institute offers the courses you want to follow.

This level of study involves narrowing down the number of subjects to about three or four. Choose carefully, thinking about how these subjects will help you with your plan or give you a good foundation for your next step. For many STEM-related careers you will need to study some of the STEM subjects at this level. Do your research so that you are not narrowing your options before you have made your decisions about possible future careers.

Extended Project Qualifications (EPQs)

EPQ is a nationally recognised qualification that can be taken by sixth-form students in England and Wales. This is an opportunity to independently explore a specific area, for example in STEM, in more detail. Students work on their own project, researching an area they are interested in and producing a report, performance or object. Keeping a log and evaluating your outcome is an important part of the process. Universities and employers value EPQs because success shows that you are interested in their STEM subject and that you have developed good organisational and project management skills.

The CyberEPQ is an example of a more structured programme that has been developed by education and industry specialists. Visit the CyberEPQ website for more information (www.cyberepq.org.uk/).

T levels

T levels are a new qualification in England, Wales and Northern Ireland. They are an alternative to A levels and more courses are being added to the list each year. During the two years of study, you will spend 80% of your time in the classroom and 20% in industry (45 days of industry placement). The courses are designed so that you learn the technical skills required by employers.

At the time of writing, there are 20 courses available including agriculture, business administration, catering, construction, creative and design, digital, education and early years, engineering and manufacturing, hair and beauty, health and science, legal, finance and marketing.

Look at your local colleges' websites to see what T level courses will be available for when you have completed your GCSEs.

Research your possible next steps carefully, as T levels are a new qualification and not all universities accept them in their entry requirements for degree courses.

Find out more about T levels at the government's T levels website (www.tlevels.gov.uk/).

BTECs

BTEC is the abbreviation for the British Technical Education Council. BTECs are vocational qualifications that are practical-based and can be studied in school or college in England, Wales and Northern Ireland. There are level 1 and 2 options (equivalent to GCSE study) or level 3 that is equivalent to A level study. From Level 3 BTEC you can progress to further study at university.

The government is planning to phase out these qualifications and replace them with T levels. (Alternative Academic Qualifications (AAQ) are also new level 3 qualifications in development, to be rolled out from September 2025. Initially, they will focus on strategically important STEM subjects.) Please research BTEC qualifications carefully as some may not be available when you come to choose your next step and some universities may not accept them for certain degree courses.

Welsh Baccalaureate

The Welsh Baccalaureate helps young people to develop the skills to 'become effective, responsible, and active citizens ready to take their place in a sustainable global society and in the workplace' (Qualifications Wales, 2023)

Level 1 and level 2 qualifications have been available for a while and level 3 Advanced Skills Baccalaureate Wales was available from September

2023. If you live in Wales, ask your teachers for more information about this qualification and how it can help with future employment.

Find out more about the Welsh Baccalaureate at the Welsh government's website (www.gov.wales/welsh-baccalaureate).

Higher National Certificates (HNC)

HNCs are taught courses that are practical-based and designed to meet the needs of employers. You can study HNCs at further education colleges and some universities. They take one year to complete but can also be studied part-time. An HNC is equivalent to the first year of an undergraduate degree. Completing this course can lead to employment, enrolling onto an HND or possibly into the second year of a degree.

Higher National Diplomas (HND)

HNDs are also taught at further education colleges and universities. As with HNCs, they are practical-based and will help you to prepare for the workplace. HNDs take two years to complete and can also be studied part-time. An HND is equivalent to the first two years of an undergraduate degree. Completing this course can lead to employment or possibly into the third year of a degree.

Foundation degrees (Fd) - A link between study progression and work

These are usually two-year courses that have been developed by employers and higher education. Following an Fd course will give you work-based technical skills that are in demand in sectors such as engineering, computing and science. Foundation degree courses are flexible and can be studied part-time while also working.

There are many HNC, HND and Fd courses available, such as in accounting, agriculture, biology, computing, sports science and different areas of engineering. Look at your local further education colleges and universities' websites for the range on offer.

If you know the subject or career area that you are interested in, HNC, HND and Fd qualifications can be a great way to learn more skills for the workplace. They can also be a good choice if you don't feel ready to enter full-time employment when you are 17/18 and are unsure about whether to apply for studying an undergraduate degree. These vocational, technical qualifications are a way of keeping your options open, as they can lead onto further study or employment.

Part-time study of a HNC, HND or Fd can be a good way to gain more qualifications while working. Some employers encourage staff to take these courses and may help with the funding costs.

> **MORE INFORMATION ABOUT . . .**
>
> *Higher Technical Qualifications - HTQ (England only)*
>
> Now, this could get a little complicated! HTQ is an 'umbrella term' that has been introduced in England and covers technical qualifications that already exist (e.g. Higher National Certificates (HNCs), Higher National Diplomas (HNDs) and Foundation degrees). HTQs are taught at colleges, universities or by training providers and are an alternative to apprenticeships. After completing an HTQ, you could go into employment or complete an undergraduate degree.
>
> New subjects are regularly being added to the list. All are approved by the Institute for Apprenticeships and Technical Education and have the HTQ quality mark logo. Visit the UK government website for more information about HTQs (www.gov.uk/government/publications/higher-technical-qualification-overview/higher-technical-qualification-an-introduction).

Foundation year

This is a taught one-year course, usually at a university, in a specific subject. Completing the year can give you access to apply for the full degree. This may be a way to get onto a degree course if your A level/Advanced Higher grades are not high enough. Have a look at the UCAS website to see the tariff points that different grades of A level, Higher and Advanced Higher are worth (www.ucas.com/undergraduate/applying-university/entry-requirements/calculate-your-ucas-tariff-points).

Undergraduate degrees

STEM degrees are highly valued by employers, as graduates develop desired skills such as numeracy, digital and problem-solving. Some employers recruit from across the STEM subjects, helping their new employees tailor their career to suit their STEM interest and abilities. (See the DLG company profile in Chapter 8 as an example.) So, don't worry if you can't decide what STEM subject to study, as having a STEM degree could help you to 'keep your options open'. As discussed

7 Study (with work experience)

in Chapters 1 and 2, look up potential employers and sectors you are interested in to see what type of graduates they employ.

Undergraduate degrees are studied at university and usually take 3 or 4 years to complete full-time. Some specialist degrees may take more than three years, for example, medicine, veterinary science and dentistry. Some further education colleges may be affiliated to a university and so able to offer degrees. There are hundreds of STEM subjects available to study at this level. In some universities it may be possible to study for a combined degree, for example, in chemistry and biochemistry. If you know what you would like to study at university, make sure you research what A level or Advanced Higher subjects are required for the course.

In Scotland, students can apply for university when they have completed Sixth Year studies (usually aged 17) and a degree takes four years. The first year of study is usually broad, covering a few related subjects, with specialising taking place in years two, three and four. Students applying from England, Wales and Northern Ireland may be able to enter into Year 2.

Applications for university are usually made through the UCAS website (www.ucas.com/), which is designed to make the process as easy as possible for school and college leavers. To apply to study for an undergraduate degree (also known as a bachelor's degree) you will need to have the minimum tariff points for that particular course and you may need qualifications in related subjects. For example, some engineering courses require A level, or Advanced Higher, Maths. Research this carefully, as there may be qualifications that some university courses do not accept as an entry requirement. At the time of writing, applying involves choosing up to five courses and producing a personal statement that explains why you'd be good for the course. You'll need to meet the course requirements and you can find these on the UCAS and university's website. (See Chapter 10 for more information about writing a personal statement.)

Sandwich undergraduate degrees

This type of degree includes an extended period (usually one year) on placement in industry. The work you will do will be related to your degree. You are usually paid a wage while on the year's placement. There are some universities that require students to undertake short periods of work placement as part of the degree. These periods of work may be unpaid.

Do your research. Things to consider are:

- How would doing a work placement benefit you?
- Can you afford to undertake unpaid work placements?

- How will you travel to the work placement each day?
- Does the university help you to find the work placement?

Read Ruaridh's story in Chapter 6 for his experience of doing a sandwich degree.

Sponsored degrees

This is when a company or organisation 'sponsors' you throughout your degree. There are lots of different schemes that exist. The support may include work placement opportunities, bursary money and payment of tuition fees.

Examples of organisations that have sponsored/scholarship degree schemes include:

- the Armed Forces (see the information box in Chapter 3)
- NHS
- Barclays
- BAE.

Some commitment towards your sponsor will be expected, for example, working for the organisation for a specified time period after completing the degree. There is strong competition for sponsored degrees, so it is important that you start researching and preparing for this pathway as early as possible.

Integrated master's degree

This is usually a four-year undergraduate course that results in a master's degree. Some students may choose to finish after three years with a bachelor's degree. In some vocational STEM subjects, this qualification can lead to industry recognition, for example, becoming a qualified pharmacist, and this may help lead to employment.

However, research integrated master's degree courses carefully, as the final year of study may not be that much different from the first three years.

Postgraduate study: Certificates and diplomas, master's degrees and doctoral degrees

After completing an undergraduate degree, some may wish to continue learning and follow a postgraduate course in a more specialised area of STEM. These can be full-time or part-time and are usually related to the first degree. This can be a way of developing your knowledge even further in your STEM subject. (Read about Molly's master's degree in media production in her career story in Chapter 2.)

Some STEM professionals choose to study for a business-related master's degree to give them more knowledge in leadership, project management or business strategy. In Chapter 9, read Doyinsola's story for an example of this.

You may think that adding two or three more years to your study is way too much! But keep an open mind, as a future employer may offer you the opportunity to take part in funded qualifications such as a master's or PhD. This could enhance your credibility within your profession.

The decision to do a master's degree could be carefully considered. Read Dionne's story in Chapter 10 for her thoughts on continuing academic study. For a contrasting story, in Chapter 9 read about what inspired Alex to study for a maths PhD.

Choosing a university

You may want to go to university straight after school/college, after a gap year or when you are older. Try to visit a few university open days so that you can make an informed choice. There are lots of things to consider when choosing a university to apply to for a STEM course:

- Does the university offer the course you want to study?
- What are the specialist teaching laboratories and workshops like?
- Does the course include opportunities to meet employers?
- What are the graduate outcomes for the course?

Websites such as Prospects Luminate (wwwluminate.prospects.ac.uk/) and the Student Room's Uni Guide (www.theuniguide.co.uk/) can help you to compare university courses.

And more generally, you could be at this institution for up to four years, so remember to consider:

- Does the university offer the social/extracurricular activities that you want to do?
- What career development does the university and course offer?
- How far away from family are you prepared to be?
- Do you want to be in a specific part of the UK or in another country?
- Do you want to be in a campus or town/city location?
- Will your travel to/from university affect your decision?
- Would you like to live in self-catering or catered accommodation?

- What does your budget stretch to (remember to include accommodation, food and day-to-day spending)?

Read Hannah's career story at the end of this chapter to find out how she chose her university.

Funding further study

16–19-year-olds can apply for funding. This differs in England, Scotland, Wales and Northern Ireland, so visit the UK government website for more information (www.gov.uk/know-when-you-can-leave-school).

Undergraduate funding exists to help with living costs. The value of the maintenance loans and grants depends on factors such as your household income, subject to be studied and whether you'll be a full or part-time student.

Tuition fees for undergraduate courses cover the cost being taught the subject. They are capped by the UK government, although the exact amount you'll pay to the university will depend on which part of the UK you are from and which part of the UK you will study in. In England, Wales and Northern Ireland all undergraduates can apply for a tuition fee loan. For Scottish students studying in Scotland, the tuition fees are covered by the Scottish government. Visit the UCAS website's money section for more information about student finance and tuition fees (www.ucas.com/money-and-student-life/money/student-finance).

Maintenance loans and tuition fee loans will need to be paid back once you are earning a salary above a threshold amount.

Having a part-time job

Many students take a part-time job while in full-time education. Reasons for this include:

- Earn money to help support themselves/family.
- Develop work-related skills.
- Develop STEM occupational skills (such as electrical engineering).
- Gain experience within a specific STEM sector.
- For social contact and enjoyment.

There are things to consider before taking on a part-time job while still in full-time education:

- Limit the hours worked so that it does not affect your study progression. Some research has shown that working more than 20 hours a week can affect academic performance.

- Find out if the employer will allow flexibility in shift patterns and number of hours worked, especially near exam times.
- If the job is for a national organisation, can you work at another local branch when you are home during the holidays?

 Fascinating fact

The music streaming platform Spotify has a 'Wrapped Team' of data scientists and engineers, animators and designers, as well as legal and marketing support to produce the annual 'Spotify Wrapped'. Each year the team improves the way the data is visualised, encouraging sharing on social media and contributing to the marketing of the company.

Conclusion

This chapter discussed different qualifications you can take with a focus on staying in education. Chapter 8 will explore going into STEM employment alongside further related training opportunities.

Remember, there is no right or wrong or one pathway, and no one can make the decision for you. Nothing is set in stone. You may decide that you want to leave education as soon as you can, but you can still choose to return to studying in the future.

The most important thing you can do for yourself is to research. Use reputable websites, speak to people who know what they are talking about and double-check the information with another source.

Sources of information:

- Careers advisers can help you to organise and develop your career ideas.
- Subject teachers can talk about their own studies at university and prior work experiences.
- Family, and other people you know, who work in the STEM career you are interested in can tell you about the pathway they followed.
- University and college websites can give you information about their courses and what it's like to study there.
- Visiting open days gives you the chance to see the college/ university and speak to people who work and study there.
- More general websites like UCAS (www.ucas.com/) and Prospects (www.prospects.ac.uk/) can give you information about courses and the careers they may lead to.

REFLECTION ABOUT CHAPTER 7

Using your reflections from Chapter 6:

- What subjects will you continue to study and why?
- Would you like formal work placements to be part of the course?
- What level of qualification would you like to achieve and why?
- Which educational institute would you like to enroll at and why?

Career story: Hannah

Primary Care Pharmacist, NHS Scotland

What is a primary care pharmacist?

We are based within GP practices and are responsible for the safe and effective prescribing of medicines within the practice. We are part of a pharmacy team that includes qualified pharmacy technicians and pharmacy support workers who work with the pharmacist. We also work with the GPs, nurses and receptionists to ensure patients are reviewed regularly, for example blood test and blood pressure monitoring, so they receive the safe and effective dose of their medication. We also handle the hospital and clinic discharge paperwork to make sure the patient gets the correct prescription. One important part of our job is to make sure the medications we prescribe are the most cost-effective, as sometimes a cheaper alternative can be appropriate.

I love helping patients. To know that a problem is fixed because I have the skills to be able to help someone is the best feeling.

Why did you want to become a pharmacist?

I loved science at school and did the three sciences at Higher and two at Advanced Higher levels. I love the medical side of things and looking after people but I am really queasy and knew I couldn't do things like taking blood. I did an online questionnaire careers activity at school, and law and pharmacy were suggested.

How did you make the decision about what to study at university?

I went to lots of university open days including the two universities in Scotland that did pharmacy (Strathclyde and Aberdeen). I did lots of research online about the courses and knew I wanted to do something science-based and patient-facing but not involving touching the patient! The open days were the most useful as you see what the campus is like, see the accommodation and the labs. You can also talk to some of the lecturers and students on the course.

Why is experience of the workplace important?

On the course, you get the chance to have work placements to find out about the different sectors a qualified pharmacist can work in. In addition to primary care, pharmacists can work in hospitals or industry and be involved with drug research and development as well as clinical trials. There are also roles in community pharmacy, such as the high street pharmacist. These placements are really important as you put the learning into practice and find out about the route you might want to go into. Some jobs are very patient-focused and others are science and industry-focused. I found the hospital work really interesting, especially working with the patients and finding out their stories.

What STEM skills do you use?

You'll usually need to have an A level or Advanced Higher in chemistry, and maths and biology are also desired. Chemistry is very important as you need to know how the drugs can interact. We use biology, for example, knowing how the enzymes in the body may affect the medication or how the kidneys breakdown the medication. We need our math skills and I definitely know my eight times tables as everything is prescribed on an eight-week cycle! Prescribing for children involves their body weight so you need your math skills to work out the dosage (milligrams per kilogram).

What changes are taking place in pharmacy training?

As of 2026, all graduating pharmacists will also be prescribers. This is another role we can do in primary and secondary care as prescribers can run clinics, see patients and prescribe medication to help control long-term conditions such as cardiovascular diseases and diabetes. The main difference between a prescribing pharmacist and a doctor is that doctors can diagnose a condition.

What advice would you give your 16-year-old self?

Keep working hard because it is worth it. It might feel that you'll never get to the end of the exams, but it is worth it. I worked really hard in school and university and I now have a really good job that I love.

8 Work (with study)

Introduction

This chapter will focus on leaving education and entering employment that includes training. This could be after GCSEs and National 4/5s or when further studies such as A levels, T levels, Highers or BTECs have been completed. If you are planning to leave school and start work before you are 18, remember to research your home nation school-leaving age (see Chapter 6).

For some, the desire, or need, to earn money is the main factor for leaving full-time education. Other students have simply had enough of being in the school/college environment and feel ready for the world of work. Some students may know exactly the STEM career they want and have researched a pathway to achieve that role through work and training.

This chapter will consider reasons for and against leaving education, different training pathways that exist across the UK and how to succeed in getting that perfect apprenticeship or traineeship. Whatever your education level, all new entrants to STEM careers will require training, so graduate training schemes will also be introduced.

Why go straight into employment?

It is important to consider the advantages and disadvantages of leaving education and going into employment:

- The initial salary may seem high, but is there the opportunity for pay progression that will allow you to live independently in the future?
- How will you travel to and from work?
- How is the apprenticeship/traineeship structured – will there be dedicated time for learning and study?
- How will your progress be monitored and how will you be assessed?
- What qualifications will you gain when you complete the training, and how transferrable are they between companies and sectors? (See Chapters 3 and 4 for more information about sectors.)
- Does the company offer ongoing professional development after you've completed the initial training?

One other thing to consider is why do you want to work for this company? It may be because it offers the training you want. Or it may be in a STEM sector/industry that you are interested in. Remember to think about what the company's values are and how they align with your own. (See Chapter 5 for more information about values.)

Traineeships (England)

This is an unpaid work placement lasting between six weeks to one year. This scheme is designed for young people (16–24 years) who have little or no experience of work and few qualifications. The course involves support with CV writing and interview preparation as well as helping participants to improve their numeracy, English and digital skills. Although there is usually no salary, travel expenses and meals may be covered.

Many STEM careers require technical skills that are gained through further learning and training. Completing a traineeship can give you the skills to move on to a STEM apprenticeship.

Please speak to your school or college careers adviser if you think you might benefit from doing a traineeship. They will help you to look for opportunities.

For more information about finding a traineeship in England, visit www.gov.uk/find-traineeship.

National Vocational Qualifications (NVQs)

National Vocational Qualifications (NVQs) are qualifications that are based on the National Occupational Standards set by the Sector Skills Councils and usually completed in the workplace. If you complete an NVQ in the workplace your employer will pay for the course costs.

Some NVQs may be delivered in schools or colleges (England, Wales and Northern Ireland) with specialist departments that have a work environment set-up, for example, construction and animal care. Courses up to and including level 3 may be government funded, depending on your previous education level.

There are NVQs from levels 1 to 7 and no age limit for enrolling. So you can find a course that suits your STEM career interest, experience and education level. The qualification is completed when you've met all of the National Occupational Standards and had your portfolio of work 'signed-off' by an assessor.

For more information, visit the NVQ page at the UCAS website (www.ucas.com/further-education/post-16-qualifications/qualifications-you-can-take/nvqs).

Scottish Vocational Qualifications (SVQs) are work-based qualifications

Scottish Vocational Qualifications (SVQs) can be taken at Scottish levels 4–11. (See Chapter 6 for tables about the education levels.) SVQs are units of study that are practical-based and usually completed in the workplace alongside your day-to-day job.

There is a range of STEM SVQs available from agriculture to computing and engineering to science. Completing an SVQ means that you have the National Occupational Standards skills, at that level, for that sector.

For more information about SVQs (Scotland only), visit www.sqa.org.uk/sqa/78913.html

For more information about NVQs and SVQs, visit www.cityandguilds.com/qualifications-and-apprenticeships/qualifications-explained/nvqs-svqs-keyskills-vocational-skillsforlife

Apprenticeships

All apprenticeships involve the same elements:

- Applications are to the employer (or recruiter), just like any other job.
- You are paid a salary (wage).
- Your employer pays the tuition fees.
- You will learn specific occupational/technical skills on the job and should have dedicated study time each week (this may be in a college or university).
- You'll quickly develop work-related skills such as teamwork and project management.
- Upon successful completion, you will have a recognised qualification (appropriate to the apprenticeship level) in that sector.

Apprenticeships are usually focused on a job role, such as mechanical engineer or cyber security analyst. They are available in just about all STEM sectors.

For some apprenticeships, you may need to be over a certain age and have the entry requirement qualifications. Apprenticeships are available across the UK, although each home nation uses different terminology to describe each level of apprenticeship. For more information, visit the home nation weblinks below:

- England (Intermediate, Advanced, Higher, Degree), www.apprenticeships.gov.uk/

- Wales (Foundation, Standard, Higher, Degree), https://careers wales.gov.wales/apprenticeships
- Northern Ireland (Level 2, Level 3, Higher), www.nidirect.gov.uk/campaigns/apprenticeships
- Scotland (Foundation, Modern, Graduate), www.apprenticeships.scot/

Read Josh's story about becoming a level 3 mechanical engineering apprentice (Chapter 3).

Degree apprenticeships

Although the terminology is different, in all four home nations there are opportunities to gain a degree while working as an apprentice. This degree qualification is no different from that gained by doing a full-time course at university. One main benefit is that the tuition fees are paid for by your employer so you don't incur the debt of going to university. There are things you should consider:

- Competition to secure a degree apprenticeship is very high.
- It may take five or more years to complete the degree.
- Work will be hard as you'll be juggling your day-to-day role with studying.
- Some studying may involve periods away, for example, attending summer school.
- During the programme, you may be offered the opportunity to study additional qualifications at a lower level that will give you further technical skills.

New degree apprenticeships are being approved each year. In September 2024, the first Medical Degree apprentices will start their course at a few selected universities. By 2031, the NHS hope to have 2,000 people training to be doctors via an apprenticeship!

To find out more about degree apprenticeships in England, Wales, Scotland and Northern Ireland, follow the apprenticeships weblinks listed above. The website Not Going to Uni also contains useful information about alternative routes into employment (www.notgoingtouni.co.uk/).

To really know what it is like to do an apprenticeship, it's best so speak to someone who has been through the experience. If you don't know a former apprentice, watch a video made by a sector organisation such as Make UK's apprenticeship videos (www.makeuk.org/future-makers/become-an-apprentice).

Read Ethan's story in Chapter 3 about doing a degree apprenticeship with Airbus.

8 Work (with study)

How to find an apprenticeship or traineeship

If you plan to follow a NVQ course at school/college, talk to your careers adviser or attend an open day event to find out more information about how to enrol.

For many work-based learning schemes (and all apprenticeships), you will need to apply to a job vacancy. You can find these vacancies on company and recruitment websites, for example, Reed (www.reed.co.uk/) and Indeed (wwwuk.indeed.com/). There are apprenticeship opportunities with national and international companies as well as local, smaller companies in your area. (See the section on small to medium enterprises (SMEs) in Chapter 3.)

Your local area may have a training provider who advertises local apprenticeship vacancies and oversees the formal learning aspects of the programme. For example, a small construction company may work with a local training provider in order to ensure all training needs meet the standards expected for that level of apprenticeship. Some training providers may be based at your local further education college.

Visit the government's apprenticeship website for more information about apprenticeship opportunities in England (www.apprenticeships.gov.uk/apprentices/browse-apprenticeships).

> **FOCUS ON AN ORGANISATION: SPECSAVERS**
>
> Specsavers are most commonly found in high streets across the UK and have three support offices in the UK (Guernsey, Nottingham, Southampton). They offer vital support and services for people with sight and hearing. Specsavers have over 30,000 employees (in the UK and globally), and the range of jobs include healthcare and business support.
>
> Early career opportunities exist for school and college leavers and graduates. Apprenticeship programmes include:
>
> - Level 2 Customer Service Practitioner
> - Level 3 Optical Assistant (eye care)
> - Level 3 Lab Technician in spectacle making
> - Level 5 Hearing Aid Dispenser (hearing care)
>
> In England, Specsavers sponsor participants through an 18-month Level 5 apprenticeship in Hearing Aid Dispensing, resulting in a Foundation degree.

> Read Bonnie's interview about training to be a Level 5 Hearing Aid Dispenser in Northern Ireland. *The HEDip is offered in Scotland for colleagues in devolved nations.*
>
> There are also roles as audiology practitioners, which is an entry-level career in audiology with Specsavers. This means that you can enter this with your school-leaving qualifications.
>
> Apprentices can be employed in the retail stores, support offices, manufacturing or distribution sites. They work towards nationally recognised qualifications while earning a wage. Personal development is really important, and apprentices can go on to clinical or management career pathways.
>
> For those with some experience in business and enterprise, there is the opportunity to become a partner and set up your own business with joint venture support from Specsavers.
>
> Visit the Specsavers careers website page to find out more (https://join.specsavers.com/uk/explore-careers/).
>
> Read Bonnie's career story at the end of this chapter.

Benefits of having a vocational qualification

People who gain work-based qualifications have successfully juggled their day-to-day job and studying. This shows employers that they have organisational skills such as time management and project management. They are able to prioritise their workload and complete tasks on time.

Having a vocational qualification also means you have the occupational skills to do the job. This is very important within STEM sectors as technical (including digital), numeracy and science skills are what businesses need their employees to have.

Is training the easy option?

Not at all!

There will be a lot of competition for these places. If you know of a local company that offers apprenticeships in a STEM career you are interested in, try to find out more about how they recruit.

You are more likely to be successful in your application if you are well-prepared and have gained relevant work experience. See Chapter 10 for more information about preparing for your STEM career and visit the

8 Work (with study)

apprenticeships website (www.apprenticeships.gov.uk/apprentices/preparing-apprenticeship).

Going into employment to gain qualifications is not easier than staying in full-time education. You'll need to be prepared to work hard, stay focused and be very organised.

Graduate schemes

Having a degree does not necessarily lead you straight into a job. You may know a lot about your subject, but your course may not have prepared you well for the workplace. Joining a graduate scheme can be a good way to develop the next stage of your career.

Most graduate programmes are for two years and are structured, allowing you to learn more about the sector and company as well as developing your technical and work-related skills. Participants usually receive career development support, for example, gaining further professional qualifications, such as Chartered status.

Read the information in Chapter 4 about NHS graduate schemes that include science and technical opportunities.

Prospects, Jobs Graduate and STEM Graduates have great information on their websites about graduate training schemes.

- www.prospects.ac.uk/careers-advice/getting-a-job/graduate-schemes
- www.jobs-graduate.co.uk/
- www.stemgraduates.co.uk/

As with apprenticeships, competition for graduate scheme jobs is very high and you will need to prepare thoroughly for the application process.

FOCUS ON AN ORGANISATION: DIRECT LINE GROUP (DLG) - DLG AUTO SERVICES

Technical graduate scheme

Any graduate with a STEM degree or HND equivalent can apply for this three-year scheme. DLG is looking for graduates who can apply the STEM skills they have learnt, so the STEM subject doesn't matter.

The scheme starts with an induction period when you get to know each other and learn a little more about DLG. Your first year will be spent at one of DLG's Body Shops and then there are four six-month rotations through different departments. This is important as you may work with those departments in the future.

The scheme enables you to acquire a good understanding of the depth and breadth of DLG AS and understand the insurance and vehicle repair business. You'll gain an additional professional qualification as a Vehicle Damage Assessor (VDA). There will also be an opportunity to develop your skills and professional interests, and to gain further qualifications.

How to apply

Application to the scheme normally opens in February and is advertised on DLG social channels and website (www.directlinegroup.co.uk/). The roles are also advertised on Handshake (https://joinhandshake.co.uk/) and through DLG partner universities.

Read the interviews of two DLG employees, Talal in Chapter 5 and Doyinsola in Chapter 9.

 Fascinating fact

Many physical features are used to sort recyclable waste at recycling facilities. Oscillating conveyor belts remove corrugated cardboard first, then rigid plastics and paper and card. Magnets remove magnetic metals and eddy currents remove other metals. Near-infrared sensors identify different types of plastics, and jets of air 'fire' them in different directions. The processes may be automated, but mechanical engineers are needed to keep things working.

Conclusion

There are benefits to entering employment as soon as you are able to, but think past the weekly wage. Training opportunities are important to help make your career sustainable, allowing you to progress and earn more money as your homelife changes and you become more independent.

Missing out full-time college or university experiences may mean you miss the chance to extend the transition period between childhood and adulthood. For some people, this time is important as they enjoy more social time and learn how to become a little more independent.

Give yourself time to research the advantages and disadvantages of staying in education or going into full-time employment. This will help you to make an informed decision that can help you to be happier and more successful with your choices. But remember, life happens and plans may have to change. See Chapter 6 for more information about how to be prepared for unplanned events.

Your teachers, friends and family may talk about academic or vocational pathways. Try not to see them as two separate routes. As Chapters 7 and 8 show, you can study while having some work experiences (e.g. T level, BTEC, HNC and HND) or you may choose to have a paid job in your chosen STEM area with elements of study (e.g. NVQ, SVQ or apprenticeship). You can also keep returning to education as your career progresses, you get older or the time is right for you.

 REFLECTION ABOUT CHAPTER 8

Whether you plan to go to college, university or straight into work, it is good to know what your choices are, so that you can make an *informed decision*.

Try to find out the following for your local area:

- Who are the local training providers (e.g. working with local companies to offer apprenticeships)?
- What training opportunities exist at your local further education college?
- Are there any STEM-related businesses near you and do they hire trainees/apprentices? (Searching local job adverts may help you to find these companies.)
- Where do local friends and family work? What STEM training opportunities exist at these companies?

Using your reflections from Chapters 6, 7 and 8:

- What do you plan to do for the next few steps in your career? (Remember that includes learning, as well as work.)
- Why do you want to follow these steps?
- What research and preparation will you need to do in order to be successful? (Chapter 10 will give you the opportunity to think about this in greater detail.)

Career story: Bonnie

Hearing Aid Dispenser Trainee, Specsavers, Northern Ireland

Describe what do you do in your role.

I am a Hear Care Assistant and training to be a Hearing Aid Dispenser, and I work between three Specsavers stores. I am about halfway through my training on a two-year Diploma in Higher Education (HEDip). When I complete, I will have a level 5 qualification. This means I'll be able to do further testing on patient's ears, prepare prescriptions for hearing aids and dispense them.

There are lots of training opportunities in this role. Four times a year I fly over to Queen Margaret University, in Edinburgh, to learn new practical skills. Last time I learnt about the in-depth testing of hearing, and I can put this into practice in the store. I always loved biology, so it is good to learn more about how the ear works. When I was younger I never thought I'd be using physics, but we learn about how sound waves work and how to assess hearing. It is interesting learning about the different types of hearing. An example is that humans are alert to different types of sound, such as screaming, as this is a survival instinct.

There is lots that I do in my day-to-day job. As well as doing hearing tests and dispensing hearing aids, I adjust patient's hearing aids. Sometimes background noise can be too loud, such as chopping vegetables, and you need to find the right balance for their prescription.

I also check the health of patients' ears. For example, you may see a mole behind their ear that they hadn't noticed. Often we are the first ones to spot sun damage people may have around their ears.

How did you find out about this job?

Not many people know about 'hear care' or audiology careers or helping patients with their hearing. My brother worked for Specsavers in a different store, and he said how great it was to work for them. I needed a career change (I was working in quality control in a food production company) and so thought I'd give it a go.

I started off as an optical assistant and did that job for two years. The director of the store highlighted career opportunities, and I said I was interested. I initially thought about optical training but my manager suggested the hearing care route. It can be harder to help someone with hearing problems. He recognised that I am a good 'people-person' and listen to what is troubling them.

What do you enjoy about your job?

I am also trained in microsuction, that is wax removal. Sometimes patients don't realise how much the wax is affecting their hearing. As soon as the wax is removed and they hear the difference, it is very gratifying. We have had some patients in tears of relief.

It is the same when fitting someone with hearing aids. Hearing loss can creep up on a person gradually and they may not realise the problem although people around them do. Family tell them to get a hearing test and they suddenly realise how much they have been missing out. Hearing loss can cause people to withdraw into themselves and affect how socially active they are. When a patient has a hearing aid fitted for the first time, they realise how much they have been shouting. It can be funny when they tell you not to shout!

I enjoy the different aspects to this job. As well as working with patients, I also prepare the hearing aids. I love building them and getting them ready for the person.

What advice would you give your 16-year-old self?

You don't necessarily have to settle on having just one career. I started out following the business studies route because of the advice I received at school. This doesn't mean I need to stay in business or go into management. Go back for more training and learn new things. Try different things and see what works out for you. I never knew anything about hear care, but I find it fascinating and a worthwhile job.

9 What do STEM job adverts tell us?

Introduction

Some people love looking on the internet at houses for sale. This might be an aspirational goal, but it is good to see what is available, where the houses are located and how much they cost. This can help you to plan for the future. It should be the same for our career. It doesn't matter how old you are or where you are up to in your education, everyone can benefit from having a little look at the adverts for dream jobs. Read Alex's career story at the end of this chapter to find out how his persistence helped him to get his dream job.

This chapter will discuss three main types of roles within STEM: technical, business support and management. This chapter will also consider why it is important for you to look at job adverts and the different places you can find them.. An example of a job advert will be examined in detail to help you understand what you can learn from them, and Chapter 10 will discuss how you can use the next few years to prepare to get that dream job.

Types of jobs

Often the job vacancies advertised can be grouped into technical, business support or managerial roles.

Technical roles

Technical roles can be found in science, technology, engineering and mathematics. You may be involved in manufacturing a product or the research and development of new products. Examples of workplaces include laboratories, offices, factories and sites such as mining, transport and construction. Examples of the products include pharmaceuticals, food, new energy technology and buildings.

Key terms in the job description can include: technician, scientist, engineer, operator, maintenance, installation, problem-solving, communication, numeracy.

Business support roles

The business support roles discussed here require STEM skills such as numeracy and digital skills. Examples of the roles include finance, IT and logistics. These business support roles can be found in STEM and non-STEM industries.

Key terms in the job description can include: Information technology, data, analysis, support, numeracy, digital, communication.

Technical and business support jobs can be for school leavers, college leavers, or graduates. Many of these jobs offer training and progression routes so that you can develop your skills and gain further qualifications. Occasionally, a person may have the opportunity to move into a role that is different from their college or degree subject (see Doyinsola's story at the end of this chapter about moving from engineering to area network controller). With enough experience, it is possible to move into a management role.

Managerial roles

As well as leading a team, managers work strategically to ensure work is done in the most efficient manner. Some managers may lead the day-to-day processes. Other management roles can involve project management for a specific project that has an endpoint.

Some of the best managers are the ones who have technical knowledge of that STEM sector and experiences of working with people. Good managers know how to support their team to thrive while delivering the best product, process or project possible.

Those just leaving school, college or university are unlikely to have the occupational knowledge and skills to gain a management job straight away. Trainee management schemes do exist in some companies; however, learning and developing your occupational knowledge first provides a solid base to move into management. Read Lauren's story in Chapter 1 to find out how she moved into management.

Many STEM companies will give their employees the opportunity to work on projects and lead sections of the work. This is a great way to develop your confidence and your teamwork skills. In turn, this can lead you to develop your leadership skills.

9 What do STEM job adverts tell us?

Key terms in the job description can include: strategic, leadership, project management, experience, communication.

Dissecting a job advert

Below is an excerpt of a job advert for a technical STEM apprenticeship. This is taken from a job advert published in January 2024 on the Indeed website (wwwuk.indeed.com/) for a Lab Scientist Apprentice at a pharmaceutical company in Cheshire. Regardless of what STEM sector you would like to work in or what subjects you are interested in, have a look at what we can learn from this advert. This 'thought experiment' can be applied to any job advert that is of interest to you. (The number in brackets highlights information that is discussed in detail below this advert.)

Lab Scientist - Degree Apprenticeship *(1)*

Starting salary - £21,500 *(2)*

About the Apprenticeship

You will join a vibrant team *(3)* that is working on the next generation of medicines and play a key role in the advancement of new laboratory techniques and technologies to support the development of new medicinal products. *(4)*

This is a 4-year fixed term contract opportunity. *(2)* You will be sponsored through your Laboratory Scientist Degree Apprenticeship including studying to complete a BSc (Hons) degree in Chemical Science. *(1)* The apprenticeship combines working, earning a full-time salary, with fully funded part-time university learning (on average 1 day/week). *(2)*

Skills and Capabilities

- Enthusiastic individuals who are keen to work within a laboratory setting and with a genuine curiosity for the science involved in creating new medicines *(4)*
- Able to act independently but also to work and contribute in a team environment *(3)*
- Good organisational skills with an excellent attention to detail *(3)*
- The ability to apply a rational approach to problem solving; making judgements based on sound reasoning, approaching problems in a logical manner *(3)*
- The ability to work on multiple tasks and understanding how to prioritise *(3)*
- Excellent verbal and written communication and interpersonal skills *(3, 5)*

Entry requirements

- 5 GCSEs (or equivalent) grades 9 to 4 including English, Maths and Science *(5)*
- To have achieved or predicted to achieve a minimum of 104 UCAS points at A-level (grades BCC), with one being Chemistry or an equivalent scientific discipline. *(5)*
- Candidates with relevant workplace experience or a BTEC Level 3 Extended Diploma in a related scientific discipline will also be considered but must provide additional details of their experience/units studied on the course as part of their application *(5)*

Successful candidates will be asked to attend a virtual Assessment day on 12th March 2024. *(6)*

Should you require any reasonable adjustments or accommodations, please let us know on your application. *(7)* We aim to set you up for success and we recognise the challenges that starting a new job can bring. We therefore offer a relocation package to help with the costs and process of moving for candidates who currently live more than 60 minutes from the site location for this Apprenticeship. *(2)*

Start date for the apprenticeship programme is **2nd September 2024** *(2)*

Where can I find out more?

Visit our website and social media, for Inclusion and Diversity information and career opportunities. *(7)*

What we can learn from the advert . . .

(1) **The job title.** This may be an obvious description of the job, as in this example. Occasionally, the job title may not mean anything to you. Let's say you're researching jobs you can do with a maths degree and *actuary* is suggested? Do you know what this means? How can you find out? Don't ignore a suggested job or role just because you don't know what it means!

(2) **These points are about money!** You need to consider where you plan to live and how you will travel to work and university. Will this salary cover this? How would you fit your plans around the start date, for example, summer job, holidays, travel? What could you do after four years? Is there the opportunity to continue working at the company? What do other apprentices say about the company? (Glassdoor is an interesting website where employees comment on the training, salary and opportunities at their company, www.glassdoor.co.uk/.)

(3) **These are all about skills**. See Chapter 5 for information about skills and why they are important in the workplace. Everyone who applies for the job will probably have the qualifications (see point 5), but they may not be able to demonstrate they have all of the skills desired by the company. By looking at job adverts a year or two before you are ready to apply, you can give yourself time to strengthen your skills and develop your weaker ones. When you come to write your applications, you can use examples from clubs, sports, hobbies and activities to demonstrate the skills you have. For example,
   - Being part of a sports team can show you work well in a group and have good interpersonal skills.
   - Making models and writing code show that you pay attention to detail.
   - Being good at chess show you are a reasoned and logical thinker.
   - Repairing simple machines demonstrates problem-solving skills.

(4) **Sector knowledge**. If you really want this job or to work for the company, you'll need do demonstrate that you're interested in the sector and know a bit about the subject. Take the opportunity to use any specialised equipment at school/college/university (e.g. in technology or science subjects). Read up on latest developments and trends in the sector. Follow the business on social media (e.g. on LinkedIn) and regularly look at their website. (This is where you can also find out about the company's values. See Chapter 5 for more information about why considering your work values is important.)

(5) **Qualifications**. To give yourself the best chance of being a successful candidate, take the time now to plan what qualifications you'll need to study and what exam grades are required. Put in the effort now in trying to succeed. Read Hannah's career story in Chapter 7 about how working hard at school will be worth it.

(6) **The interview process.** If your application is successful, you'll be invited to attend the next stage of the interview process. This could be an online assessment, an online or telephone interview, a face-to-face assessment event and finally a face-to-face interview. These events can sound daunting, but try to prepare as much as you can beforehand. Take part in mock-assessment centre events, ask a senior teacher/lecturer to give you a mock interview, get yourself comfortable with using the telephone and online meeting platforms *with the camera on!*

(7) **Inclusion and diversity** are important for every organisation, and many companies are working hard to ensure everyone feels valued. (See Chapter 2 for more information.) Could you be comfortable and happy working for this company? Before you apply, look at what their website says and read reviews on a reputable website

such as Glassdoor. If you have any requirements to help you perform at your best during the interview process, then let them know. If you have any questions about the recruitment process and working for the business, get in touch with the contact in the job advert.

The advert above is for a scientific role. Both science and careers education have changed a lot during the last two decades and one of the people who has been part of this development is Sir John Holman.

 HERO IN STEM: PROFESSOR SIR JOHN HOLMAN

You may never have heard of Sir John Holman, but if you live in England, it's likely that he's had an impact on your science education and the career learning you do at school and college! Sir John is a respected researcher and adviser of science and careers education.

Sir John studied natural sciences at Cambridge, gaining a Bachelor of Arts degree, and then started his career in chemistry teaching. Gaining funding, he was able to combine teaching with science education research at the University of York. Sir John has been involved in many initiatives that shape current science teaching, especially chemistry, such as Salter's Advanced Chemistry course and the formation of the National STEM Centre in York (where many school laboratory technicians and science teachers have received training).

Careers education in schools and colleges has undergone major changes during the last ten years, and you may have heard of the Gatsby Benchmarks. Sir John led the development of this framework that aims to ensure every young person in England has access to careers guidance, information about academic and vocational options and the opportunity to meet employers through experiences of workplace and speakers coming into school. The work that Sir John led has also helped to shape careers education in Scotland, Wales and Northern Ireland.

Where can you find STEM job adverts?

You can find STEM jobs advertised locally or nationally, and the internet is the best source.

- Company websites (remember to look for STEM roles, such as IT and financial, in non-STEM businesses, too)
- Recruitment websites such as STEM Recruitment www stemrecruitment.com/), Reed (www.reed.co.uk/), Indeed wwwuk. indeed.com/) and Gradcracker (www.gradcracker.com/).
- Local websites, such as training providers and councils.

There are other ways to find out about who is hiring and how they recruit. You may not be ready to apply yet, but just as with reading an advert, you can learn something about the recruitment process by attending an event:

- Job fairs in local towns and cities
- Job fairs in colleges and universities, for example graduate recruitment
- Hiring events, with one company that has multiple roles to fill.

But I'm not ready for a job yet!

You may not know what type of career pathway you would like or what sector you want to work in. However, looking at job adverts is useful research. It can reinforce that skills are important to employers. So you can start to plan how you can develop yours and demonstrate this. You'll find the same skills keep cropping up in job adverts (such as creativity, teamwork, organisation and problem-solving). (Read Catriona's view on problem-solving in her career story in Chapter 4.)

On-site, remote or hybrid?

The world of work has changed so much recently, accelerated by the pandemic. Perhaps your lessons and lectures changed to online for a while? The workplace has kept some of these changes. Many people whose job involves some element of computer work (this includes apprentices, trainees and graduates) may have part of their week working at home.

Think about how you work best. Do you thrive in company? Will you get lonely on your own? Need a bit of peace and quiet to get stuff done? Enjoy bouncing ideas off others? Will you work best on-site or remotely at home? Will you have the space at home to work? Are you naturally quiet or shy and would benefit from the structure of regularly working in an office or on-site? Perhaps a bit of both is best for you and you'd embrace a job that is hybrid? Know the positives and negatives of home-working and be prepared to discuss this at an interview so that you and your company create the conditions for you to succeed in your new role.

You may hear about the importance of being adaptable and agile in the workplace. But what does this actually mean? This means that you can deal with changing circumstances and continue to thrive. Having certain behaviours and characteristics can help you to do this, such as being curious, optimistic, resilient and flexible.

Read Talal's story in Chapter 5 to find out how he had to adapt to home-working during the pandemic.

 Fascinating fact

121.6 grams of dust collected from asteroid Bennu in the OSIRIS-Rex space mission landed on Earth in 2023. The asteroid is 4.6 billion years old, about the same age as the solar system. Researchers, including astrophysicists and analytical chemists, at the UK's Natural History Museum have been given 0.1g to analyse to learn more about the formation of the solar system.

Conclusion

This chapter focused on different types of STEM roles (technical, business support and managerial) and how looking at job adverts can help you to prepare for applying for a job.

Looking at the vacancies advertised for the types of job you aspire to can help you to decide the qualifications and training routes you may need to follow. You can begin to plan the steps you'll need to take to be the ideal candidate.

Although it is beneficial to plan, plans may need to change due to circumstances outside of your control. Also, remember to keep an open mind about the type of STEM career you may have. You could end up seeing an opportunity for a job that you had not come across before.

Chapter 10 will discuss how you can prepare during the next few years to be the best person for the job.

REFLECTION ABOUT CHAPTER 9

Use the information above to find examples of STEM jobs that you are interested in. Things to consider are:

- adverts for vacancies in your local and wider areas.
- adverts from a range of companies: SMEs (see Chapter 3), national and global.

What would you need to do over the next few years to be able to apply for that 'dream' job?

Questions to consider:

- What qualifications will you need and what are the different ways you can get them? (See Chapters 7 and 8 for more information.)
- How can you improve your understanding of the sector/type of job role?
- How would you gain the skills desired?
- What experiences could help, for example, work experience, volunteering, extracurricular clubs, sports and hobbies, paid work and family responsibilities?
- Who can you talk to for advice and support?

Your reflections from this chapter will be useful in Chapter 10.

Career story: Doyinsola

Area Network Controller, Motor Fulfilment Business Area, Direct Line Group (DLG)

What is your job?

I work with the garages that do the motor repairs for DLG customers who have had a car accident. I use data to help the garages perform efficiently to DLG standards while also making sure that the customers are satisfied. I need to be good at developing and maintaining relationships, so that both the garages and my managers are satisfied.

What qualifications did you do when you were younger?

I moved to South Wales from Nigeria with my family when I was eight years old. I always enjoyed learning about how things worked and how to solve problems. I also enjoyed Performing Arts and so in the sixth form I studied science, engineering and Performing Arts A levels. After school I studied petroleum engineering at Portsmouth University. I then stayed at the university and did a master's degree in project management.

What was university like?

I left Wales when I was 18 and met so many people at university from different parts of the UK and the world, including Nigeria. I did lots of activities while I was there and was involved in the society of street dance and, as I'm a man of faith, I was involved in the campus-led fellowship. I also met my wife at university. When I finished my degree I was able to stay at the university and keep my social network and enroll on the master's course. The course had lots of students, including international students, from different disciplines, such as accounting and law. It was good to hear their different perspectives about their careers as some already had set plans and others had no idea about what they would do next. I was in between, I knew I had technical skills from my degree and wanted to learn the 'soft' skills about business and management.

How did you get the technical graduate job at DLG?

I used the careers support at the university to develop my CV and find out about how to do job applications. I also signed up on websites, such as Grad Cracker (www.gradcracker.com/) to find out about STEM graduate opportunities. It was nice to receive the emails about options I could apply for, but I was still finishing my Master's dissertation at the time. I didn't manage to get a grad job straight away, so took a charity-work job and kept searching. I saw the DLG job and wasn't sure exactly what the job description meant, but they wanted a STEM graduate and I had the technical skills, so I applied. The interview process had lots of stages, including on the phone and in-person. It was arduous but worth it as they got to find out more about me, not just what was on my CV.

What advice would you give about having a plan?

It can be difficult to trying to find your way, but keep true to yourself and your values. My faith helps me not to worry about tomorrow but to take it one step at a time. I did know that I wanted my job to allow me to live back in South Wales, to be nearer family, and that opportunity came up because of hybrid working. I've never chased a particular job as you may get the opportunity to do a job that you didn't know exists. I want to do my best, keep developing my skills and doing purposeful work that helps people and the business. I like going to bed knowing 'that happened because of me'.

Career story: Alex

Fast Particle Modeller, UK Atomic Energy Authority, Oxford

Describe your job

I work for the UK Atomic Energy Authority (UKAEA) on fusion energy research. All the nuclear power plants that exist right now are fission power plants. They take heavy elements, break them down to make lighter elements and this releases energy. But fusion is the other way around, taking light elements to make heavier elements, releasing energy. We want to fuse together two heavy types (or isotopes) of hydrogen, deuterium and tritium.

My main task is helping to design a future fusion power plant called STEP* which we aim to build in the 2040s. It will be the UK's first fusion power plant that will put energy into the National Grid. This is a massive project and my role is specifically to model the helium ions (or so-called alpha particles) released from the fusion process and try to predict what impact they have on the machine, thus informing the design process. If I calculate the alpha particles will hit the STEP wall in large amounts, we need to change the design to reduce the heat load on the wall from the alpha particles.

Most of my work is done on the computer, using simulations to do the modelling. But I also work with colleagues, discussing the work and having meetings. Although I am a modeller, I've found that I also enjoy computing and learning about coding. I like to make the code that the modellers then use.

*For more information about fusion energy research and STEP, visit https://step.ukaea.uk/.

What subjects did you study?

When I was doing my A levels I was pretty sure I wanted to do physics at university. I did physics, maths, further maths, chemistry and geology at school.

I didn't think beyond university, I just knew I wanted to do physics. I went to St Andrews and started a physics degree but switched to maths after the first year. When I was in school I thought maths was too abstract and a bit boring. But at university, I found physics more difficult, especially the lab practicals, and found the theoretical work and the maths easier. Experimental physics and doing lab work just didn't suit my personality.

I was an English student going to a Scottish university where degree courses are four years. But because of what I had studied in my A levels, I was able to enter Year 2. This is called 'direct entry'. It was quite difficult skipping a year because in theory you have covered the stuff, but I found it wasn't really like that and I had lots to catch up on. My first year at uni was the hardest; it was easier when I switched to maths.

What opportunities did higher education offer?

In my penultimate year at St Andrews, I did an internship with the solar physics department. This was my first exposure to research. This is completely different from undergrad work, and I really enjoyed it.

So I then decided to do a PhD in solar physics with the same professor who supervised the internship. This took me three and a half years and was funded by Science and Technologies Facilities Council.

What happened next?

I read about the UKAEA in a solar physics newsletter and found that I could apply for a fusion research job as, like solar physics, it is just a type of plasma physics. I applied to work at UKAEA three times but, although I got to interview, I was rejected for the first two jobs I applied for as I wasn't the best person for the role. I knew the company would be a good fit for me, and they encouraged me to apply again if I saw other roles that I thought I would suit. I eventually started working for them in September 2021.

10 Preparation and planning for a STEM career

Introduction

This chapter will help you to pull together the ideas from the book so far so that you can make your next moves:

- Looking back over Chapters 1 and 2, are you happy that you understand what a career in STEM means and whether this could be for you?
- From the information in Chapters 3 and 4, are there STEM sectors, and perhaps career pathways, that you are interested in?
- Looking back at Chapters 5 and 9, do you have an idea of your strengths and skills that you need to develop?
- Are you aware of the different pathways into work discussed in Chapters 6, 7 and 8? And do you understand that there is no one definitive route for you into employment and that both education and vocational training are important, and you can do a mixture of both.

You are now ready to plan your next moves . . .

Getting started

Have you ever tried to do something but failed?

Being motivated is the starting point. Knowing *why* you want to do that thing can help you to succeed. And this means **setting a goal**. For example, saving money *so you can buy a new game*, learning to play a guitar *so you can join friends in a band* or passing a driving theory test *so that you can take the driving practical test*. In each of these examples, there is a tangible goal that would keep you motivated.

Here are some examples of the different types of goals you can set for yourself on your 'journey' to a STEM career:

- short-term goals: passing subject tests, joining a STEM club, gaining work experience
- medium term goals: developing your skills, passing exams, learning to drive, getting to university
- long-term goals: getting a job in a specific industry, for example, nuclear, or getting on the pathway for a particular career, for example, medical doctor or owning your own home.

Planning

Creating an action plan involves listing actions and including the activities with set dates you want to achieve them by.

What could you plan for?

- developing a skill
- passing a qualification
- choosing a college or university course to follow
- applying for an apprenticeship
- aiming for a particular university
- researching a particular career.

Here is one way to write an action plan. Use this example as a guide to write yours.

Goal	E.g. getting particular grade in a subject exam	
Why do I want to do this? (This is your motivation.)	E.g. so that I have the required qualifications to apply for a specific course	
What I must do in order to succeed?	**List of actions**	**Date to complete**
	Attend all lessons/lectures	Daily
	Complete and hand in all homework	Weekly
	Attend the revision sessions	Weekly
	Get pass, or above, in each test	Monthly

You may have heard of SMART action planning. This table can help you to gather your thoughts for your action planning:

Specific	What is my goal?
Measurable	How will I know when each part has been achieved?
Achievable	Can I do it? Who do I need to work with to help me?
Relevant	Why is this goal important?
Timed	Set dates for when actions are to be completed.

Personal statements and CVs

Personal statements are usually written when you are applying for courses. When I was training to be a careers adviser, my mentor taught me the ABC of writing a good personal statement: Activity, Benefit, Course. For any activity you have done that you write about in your personal statement, describe what the benefit was to you and how this can be applied to the course. Here is an example:

- Activity: volunteering for St John's Ambulance
- Benefit: learning first aid and skills to work calmly under stressful situations
- Course: for example, applying this to a paramedic science degree; practical experience of giving first-aid support during community events.

Your CV is a brief document (one to two pages long) that you can use to demonstrate who you are, what you've done and what you're good at. This book won't tell you how to write your CV, but there are plenty of resources to help you with that task! For example, the book *You're Hired! Standout CVs* by Corinne Mills (Trotman) will guide you through the process of writing your CV. The Indeed website (wwwuk.indeed.com/career-advice/cvs-cover-letters/stem-skills) has tips about the STEM skills to include on a CV (Chapter 5 has information about the different ways to group skills) and the Reed website has CV templates (www.reed.co.uk/career-advice/cvs/cv-templates/).

You may get the opportunity in school or college to write your CV and receive feedback from a careers adviser or even employers. Make the most of this, as it will make writing your CV for an actual job much, much easier.

Sources of STEM and career-related information

You are reading this book so you are already a careers researcher. Teachers, family and careers advisers can tell you things, give you guidance and encouragement. But it is up to you to do enough research so that you make informed decisions.

Do you believe everything that you are told? What helps you to decide if a source of information is truthful? How can you double-check that a fact is correct? It is important that you learn how to critically analyse and verify information from different sources (also see Chapter 5, digital literacy).

Examples of STEM career websites to explore:
- Creative industries /www.meet-eric.com/
- Biology www.rsb.org.uk/careers-and-cpd/careers
- Chemistry www.edu.rsc.org/future-in-chemistry
- Physics www.iop.org/careers-physics
- Information technology www.bcs.org/it-careers/
- Engineering www.theiet.org/career/
- Maths www.mathscareers.org.uk/
- STEM www,stemettes.org/young-people/

Examples of other career websites to explore:
- Prospects www.prospects.ac.uk/
- Youth Employment UK www.youthemployment.org.uk/
- Career Pilot www.careerpilot.org.uk/
- My World of Work (Scotland) www.myworldofwork.co.uk/
- Careers Wales www.careerswales.gov.wales/
- Bright Network www.brightnetwork.co.uk/career-path-guides/

People who can support you

Often, you just need to discuss ideas with another person who can ask you questions and help you to gather your thoughts.

There are people you can talk to about ideas for your future career plans:

- **Careers adviser:** Every young person in the UK is entitled to speak with a careers adviser. A careers adviser won't tell you what job to apply for or which course to take. These are trained professionals who can encourage and guide you to plan your next steps. Always take up the offer of an appointment with a careers adviser in your school or college, even if you know what you want to do. They will help you to explore the options so that you are making an informed decision. If you are home-schooled, look at Chapter 11 for ideas about how you and your family can access career guidance support.
- **Teachers:** They may not have qualifications in careers guidance, but your teachers can give you information as they know their subject and its importance in the world outside of education. Your teachers may be able to help you access college and university events. They can also form links with local employers who may be able to visit the classroom. You can ask your teachers about former students (alumni), as they are a great source of careers information and inspiration.
- **Family and friends:** This is the group of people you are most likely to turn to for advice. Our close social network can offer us encouragement and support. They also know you best. So when you are thinking about your skills and values (see Chapter 5), ask your family and friends what they think are your strengths and areas

to develop. (Read Bonnie's story in Chapter 8 to find out how her brother helped her when she needed a change of career direction.)

Do remember, your family and friends are unlikely to be experts in every STEM sector and industry. Sometimes people can have *'job fog'* when they don't understand a career pathway. Just as we can't see clearly when the weather is foggy, when someone has job fog, they may not be able to guide you as best as they might want to. This is the time to talk to careers advisers and teachers, and to research carefully on the internet.

- **On social media:** The various platforms such as Instagram, TikTok and YouTube are brilliant sources of careers information. If you want to know how to become a games designer, search on these platforms for 'first-person' career stories and you will get some great advice. Remember, just as with all sources of information, think critically about what your search produces, what you hear and verify the facts.

Helping yourself

There are lots of things you can do during the next few years to help you to prepare for your STEM career:

- Keep up to date with the developments and issues related to the areas of STEM you are interested in (see Chapter 1 for more information).
- Read about people you can identify with who have a STEM career (see Chapter 2 for more ideas). If possible, use your network to help you to meet and talk to someone doing a job in an area you are interested in. (See below for networking ideas.)
- Take part in activities that can help you to develop skills and knowledge in the area you are interested in, for example, take part in any STEM clubs in school and college. Try to gain experiences of the workplace (see the following pages for more ideas).
- If you know what career you would like, do your research so that you choose to study subjects/qualifications that are required (see Chapters 6, 7 and 8 for more information).

You are the most important person who can control what you do in your life. Try to develop behaviours and skills that will help you to make the most of opportunities that arise. Be curious, flexible, persistent and optimistic. See Chapter 6 for more information about how these behaviours can help you cope with unplanned events.

Sometimes life throws us a curveball and things don't go the way we planned. You may have heard the expression that someone was 'born lucky'. Perhaps they were just more open to opportunities and took the odd risk, tried new things and remained positive.

In Chapter 9 read how Alex changed subjects at university when things didn't turn out as planned.

Work experience (or *experience of the workplace*)

One of the best ways to find out if something is for you is to go and see for yourself. If you can get into the workplace you can ask questions, find out what a job (or company or sector) is really like.

Your school/college will probably encourage you to do a week's work experience that you may have to arrange yourself. This is a brilliant opportunity to find out more about a STEM area or career pathway. Be open-minded when you try to organise this and remember the four behaviours listed above.

Let's take cyber security as an example (although this could work for any career idea). You may not have any contacts who work in a company that specialises in this. But every organisation needs to protect their data and information systems. Perhaps a family member or friend works in a business that could offer you a week's work experience and you could focus on finding out how they store information, keep data secure and promote digital literacy within their staff. Part of your week may involve 'duller' stuff, such as seeing how an office is run, but remember, flexibility is important. When you apply for further qualifications, you can then use the ABC technique (see the start of this chapter) on your personal statement.

Read Dionne's story at the end of this chapter to find out how doing work experience helped her to realise her dream of becoming an engineer. Dionne used her engineering teacher's connections to help her find work experience. Perhaps there are people at your school/college who can help you?

This section heading also includes **experience of the workplace.** If a company you approach doesn't have the capacity to offer you a whole week, politely be persistent and ask if you could just join them for a day to find out about the career pathway/sector you are interested in. Take any opportunities offered in school, or by family members, to visit any type of workplace (office, lab, factory, site) and ask questions:

- What is it like to work here?
- What qualifications did you do to get this job?
- How do you use science, maths, technology in your job?
- What training and development opportunities do you have?

There are national organisations that help young people to have experiences of the workplace. Competition for the places can be tough, so you'll need to prepare your application carefully. Since the pandemic, many organisations offer **virtual work experience**. This can be a good way to learn about working online, as many professional jobs can have some element of home-working nowadays. If possible, plan a few different *experiences of the workplace*, as this will help you to gather lots of information to help you with your decision-making.

Visit these weblinks for examples of organisations that may have work experience (or experience of the workplace) suggestions and opportunities.

- www.nationalcareers.service.gov.uk/careers-advice/types-of-work-experience
- www.uk.indeed.com/career-advice/finding-a-job/how-to-find-work-experience-year-12
- www.destinationstem.org.uk/opportunities
- www.springpod.com/virtual-work-experience/search
- www.scienceoxford.com/schools/secondary-schools/stem-work-experience/
- www.stemcells.cam.ac.uk/join-us/students/work-experience
- www.in2scienceuk.org/our-programmes/in2stem/

Is it all down to you? (Networking)

We all need help and support from family, friends and staff who work in schools and colleges. These connections may know someone who can help you to gain experience of a workplace you are interested in. You need to ask for help.

How do you ask? Here is an example. If you are considering a career in architecture, but don't know an architect, how do you find out more about this career? Find out which of your connections has recently had an extension built (perhaps it was your school!). Ask them if they would contact the architect to find out if they can share your contact details so that you can enquire about work experience (always use a school or college email address).

Perhaps you don't think you have any connections with a sector/career you are interested in. Research local companies in that sector. Double-check your connections (i.e. your network) to find out if anyone works there or knows the company. They might not do a STEM job but could still work there and be able to help you. If you do have a connection, ask them if they can introduce you to a person in the company who might be able to help. If you have no connections, use the website to find the name of the business owner and write to them explaining that you would like to find out more about the industry and ask if they would be able to help. (Again, always use a school or college email address.)

Do use your family, friends or teachers to help you to write a good, well-structured letter or email to the company that clearly states what help you require (e.g. work experience, a visit, or to talk to someone who does a particular job) and explain why this is important to you.

Everything described in this section is called networking. As your career develops you may hear this phrase. It is good practice to begin to develop your network and expand your contacts as you move into further education and employment. You never know who might be able to help you in the future *and* who you might be able to help!

Read Ethan's story (Chapter 3) to find out how winning a competition he entered at school gained him a work experience visit with Transport for London.

Reviewing and adjusting plans

Chapter 5 focused on skills and four were linked to self-awareness:

1. Reflecting on progress: How are things going?
2. Goal setting: Do you know what you are working towards?
3. Adaptability: Are you having to adjust how you do things?
4. Autonomy: Can you do this or do you need a bit of support?

Every few months check how your plans are going and think about any adjustments you might need to make in order to achieve your goals. Be aware, though, is it bad planning or lack of motivation that is affecting your progress? What can you do to get back on track?

 Fascinating fact

In 2022, amateur mathematician David Smith from Yorkshire 'discovered' a 13-sided shape, nicknamed 'the hat'. However you arrange the tiles of the shape, the pattern will never repeat. Creative architects and designers may now incorporate this into their ideas. Biscuit cutters are already available for bakers!

Download 'the hat' at www.printables.com/model/435459-aperiodic-monotile-polykite-the-hat-einstein-multi

Conclusion

Congratulations! You have nearly finished reading this book. But, in reality, you've probably been skimming through and picking out the bits that interested you.

10 Preparation and planning for a STEM career

There are a huge variety of careers within STEM and STEM jobs within non-STEM sectors, such as retail and hospitality. Sometimes the creative industries are overlooked when we talk about STEM careers. But think about music, movies, TV, podcasts, the media, theatre, gaming, fashion and art. Advances in technology are providing unique opportunities for creativity and innovation.

A career in STEM can be for anyone. It doesn't matter about your social background or your academic ability, although you do need to be realistic. Becoming a doctor or a veterinary surgeon requires certain qualifications. If you don't think you can achieve these, what alternative pathways are there that are closely related to that career? Are there any career stories in this book that resonated with you? Can you find a role model who can inspire you?

Exciting developments in technology, such as automation and artificial intelligence, are leading to new opportunities. Be curious and optimistic and open to learning new skills, whatever career pathway you have.

Finally, you are the most important person to help you to reach your dream career. You might not be in control of everything in your life, but you are in control of how you react to tasks, people and the events you are part of. Developing good career management skills and positive behaviours can help you to set goals, plan how to achieve them and take the first steps on your STEM career pathway.

GOOD LUCK!

FOR THE REFLECTION ABOUT CHAPTER 10

1. Use the planning section in this chapter to write an action plan for one short-term goal.

Or

2. Use the information to:

Practise writing an 'ABC' paragraph for a college or university course you are interested in.

- Activity: What have you done?
- Benefit: How did this help you to develop?
- Course: How can this skill/knowledge apply to the course?

Ask a teacher or careers adviser to read your action plan or paragraph and give you feedback.

Career story: Dionne

Process Engineer, Automotive Manufacturing, Jaguar Land Rover

Can you describe your role?

I currently work in a small team on the new Range Rover models. We work with the designers and the manufacturing line to make sure that the new design can be built efficiently in the production line. Jaguar Land Rover is a huge organisation, and there are different teams working on specific parts of the car, for example, the bonnet, bodywork or windows.

Describe your day-to-day job

Some days I work in the office on standard operating procedures and CAD design. I often then go onto the shop floor (where the production line is) to show the engineer how they are going to use the standard operating procedure to build that bit of the car. When I am on the shop floor I wear High Vis and old work trousers as it can get greasy.

How did you get into engineering?

I went to an engineering high school and I did GCSEs and a Young Apprenticeship Scheme in Engineering. I went to Wolverhampton College one day a week and learned engineering skills such as how to use machinery, including the drill and lathe, and how to cut metal. We also learnt how to read engineering drawings and draw technical specifications. I really enjoyed this and looked forward to these days in college.

I knew I wanted to carry on studying engineering but I favoured the design side. So when I started my A levels, I thought I'd focus on maths, product design and art. I felt I already had the basic engineering skills, so initially I chose not to take A level Engineering, but my teacher persuaded me to and he was so right. He encouraged us to take up work experience opportunities in the holidays with different companies.

How did work experience help you?

I used my holidays to gain so much work experience. By the time I was writing my personal statement for the UCAS application I had lots of experiences and knowledge to write about that I actually got a scholarship to Loughborough University. I studied for a Design with Engineering Materials degree. I was able to bring my love of design and engineering together. After I graduated I went to a talk about getting more women into the engineering sector, and I was able to share some of my experiences. It was at this event that I met the person who was to become my mentor.

How has having a mentor helped you?

At that event he asked a question that I wasn't confident enough to answer in public but I spoke to him at the end of the talk and he invited me to another event in London. Since then he has been with me every step of the way. He helps me with my job applications and visits me in my job. He's even met my young daughter. He has so much engineering experience, including consultancy, that I trust his guidance for my own career progression.

What is exciting about being an engineer?

In one of my previous jobs I came into work one morning and found out that I would be travelling to India the next week to work on a project. I love the fact that I can say my job is never boring. No day is the same when you are an engineer. You can come into work one morning and have to solve the problems that occurred on the night shift. Being an engineer always involves teamwork. It's never just about you, the process is smoother if you have as many eyes as possible on the problem.

What are your long-term goals?

I'd like to become a chartered engineer, but this would require a master's degree and I'm not ready to go back to studying yet. For now, my mentor is encouraging me to become a certified incorporated engineer (IENG) as this suits me as it's for a technical, hands-on engineer. I'd really love to become a consultant one day, but I know I need a bit more experience before I can be sent into a company and fix the problems they are dealing with. This is my dream for in ten years.

11 For parents/carers

Introduction

Thank you for picking up this book. Perhaps you are considering buying it for a young person, or your child or relative may have brought it home from school.

This chapter is written for parents and those who look after or support young people in the readership age group (approximately 11–18 years). I hope the book will be useful to you and your family, regardless of which UK home nation you live in. STEM sectors that exist across the UK are discussed, and the book includes interviews with people in their early STEM career from Wales, Scotland, Northern Ireland and England.

The definition of career that I use is an individual's 'journey through life, learning and work' (Andrews and Hooley, 2022). This is taken from the *Careers Leader Handbook*, also published by Trotman, which is used by most schools in England.

I hope you find the knowledge and experiences I have gained through over 20 years of teaching science, running STEM clubs, working with employers and studying for a master's degree in careers education useful.

Young people managing their career

Your young person has already started to manage their career. They have chosen subjects to study and may take part in various activities and hobbies. Maybe they volunteer or have a part-time job. These all contribute to their career development.

Family and friends are vital in supporting and influencing young people, as are social media influencers, celebrities and others in the public eye. It is important that young people receive a balanced range of information and advice and that they know how to verify information and weigh up the alternatives so that they make good, informed decisions.

A word about motivation: Your young person can only effectively plan their next steps when they feel motivated to head towards a goal.

That goal doesn't need to be a defined career or job that they want. It could be passing a maths exam, learning a new hobby, taking part in volunteering or getting experience in the workplace. Encouraging these small steps is important to help them develop their self-confidence. When young people feel confident in their ability, they'll feel able to set bigger goals.

It is OK for your young person not to know what they want to do, but it is not OK for them to do nothing. Chapters 1 and 2 offer guidance about exploring interests and linking these to career research.

Perhaps your young person is 'dead-set' on a career plan. This is OK too, but encourage them to research the options so that they are making an *informed* decision. They may still stick with their original idea, but at least they have explored what the alternatives are and why their preferred choice is the best one for them.

Job fog

Do you ever get into a career conversation with your young person but don't really know what type of job or industry they are talking about? You are not alone. In an article in May 2022, *The Guardian* used the phrase 'job fog' to describe how parents feel when their children want to talk about unusual jobs and careers, often at the cutting edge of technology and in the creative industries (www.theguardian.com/money/2022/may/19/how-is-that-a-real-job-parents-struggle-to-keep-up-with-childrens-career-options).

There are 60 jobs discussed in relation to specific sectors in Chapters 3 and 4, and at the end of this book, we have an A–Z list of even more STEM jobs. You or I can't be expected to know about every job, but we should be able to help our young person explore the areas that interest them. Encourage them to use reputable internet sites to verify career ideas they may have gleaned from other sources. A good place to start is the Prospects Job Profiles website. This is written by career professionals and contains lots of hints, tips and weblinks that you and your young person will find useful (www.prospects.ac.uk/job-profiles).

Using this book

This book may be useful for young people who already know they want a career in STEM and for those who are not sure about what they want to do.

The introduction for each chapter explains what will be covered. Each subsection is chunked with clear headings so that those dipping into the book can quickly find information of interest. The conclusion repeats the main ideas discussed, explaining why these are important for the reader considering a career in STEM. Each chapter ends with suggested reflection activities to help the young person use the ideas discussed in the chapter to further develop their career management skills.

An overview of the content of each chapter is given in Chapter 1. There are three additional features throughout the book: Career stories, Heroes of STEM and Fascinating facts. Further information about these features is given in the section 'How to use this book' at the start of the book.

Supporting your young person

I once read a story about a movie set designer who was inspired as a child by a photograph of a TV show scene in a magazine. He wondered how they could build a medieval house and went on to study carpentry before gaining employment with a production studio. You really never know when a key moment can occur that turns a young person's life!

Ways you can support your young person:

- It is *never* too early to start career conversations. Talk about the jobs you and other members of the family have. Who works in a STEM sector, and what do they do? (Chapters 3 and 4 will be useful to help catagorise the different STEM sectors.) The work may not seem directly related to STEM, but using computers and other technology and being able to handle data are part of just about every job.
- Encourage your young person to develop good numeracy, written and verbal communication skills and digital skills. Chapters 1 and 5 include lots of information to help with this. The workplace has always undergone change; think about the invention of the printing press and the industrial revolution! Today's changes are mainly based on digital technology. Learning and professional development never end, so, as appropriate, encourage your child to explore new technologies and learn new digital skills such as coding and working with data. This will give them the right mindset for flourishing at work and adapting to change.
- Use your network. You have connections (or your connections may have connections!) to the world of work that can help your child. Perhaps they can host a visit, arrange for your child to shadow a member of staff for a day or even offer a few days' work experience. You may need to 'think outside the box' and Chapter 10 offers more guidance.

- Encourage your young person to set realistic goals, plan small and achievable steps, review progress and amend as necessary in order to achieve the end goal. Chapter 10 offers advice to help with this, including how to sensibly engage with social media and avoid being stuck in an 'echo chamber'.

For more information for parents, visit
https://notgoingtouni.co.uk/parents
www.talkingfutures.org.uk/

Home-educated?

Children in state schools are entitled to careers guidance from an appropriately trained careers adviser. They also receive a careers education programme that includes the opportunity to meet a variety of employers; information about apprenticeships, further and higher education; experiences of the workplace and local labour market information.

Try to give your child similar information to help them make informed decisions that will positively affect their long-term career outcomes. Ask your local education authority for access to programmes and events to support career learning for home-educated young people. Contact the careers advisers at your local further education colleges, as they can also give advice about study routes and career pathways. If you are in England, you can also contact your local careers hub, which works in partnership with the Careers and Enterprise Company. The Youth Employment UK website also has career resources for those who are home-educated (www.youthemployment.org.uk/home-education-uk-resources/) as does Careers Connect (https://careerconnect.org.uk/home-educated-young-people/).

For specific home nations support, visit:

- Wales https://careerswales.gov.wales/parents/your-home-educated-child
- Scotland www.skillsdevelopmentscotland.co.uk/what-we-do/scotlands-careers-services
- Northern Ireland www.nidirect.gov.uk/articles/helping-young-people-career-choices

Conclusion

It is important for you to know that a career in STEM can be for anyone, regardless of gender, academic ability, physical ability or background.

Your young person may benefit from having a role model that they can relate to. You can help them find a suitable one by contacting their school, asking friends and family or through reputable sector websites on the internet.

There are a huge variety of careers within STEM and exciting developments in technology are leading to new opportunities. However, our careers are not linear, and studying a subject at college or university doesn't necessarily mean that it is what one will do in their future career. Many of the interviews in this book show how versatile a STEM qualification is in leading to other job opportunities.

Encourage your child to develop behaviours such as curiosity, optimism, flexibility and persistence. This will help them to be confident enough to take opportunities, overcome difficulties and make the most of unplanned events. Support your young person in developing their agency so that they are able to make informed decisions in their first steps towards a STEM career.

12 For teachers

Introduction

This chapter is written for teachers, explaining how you can use the book and the accompanying resources on the publisher's website, to bring more career learning into your classroom.

You'll already be doing this in your teaching, but I hope that this book gives you a few more ideas.

I was a science teacher for 24 years and have an MA in careers education and coaching. I love the intersection of education and career development. Teachers are in the unique position to support and encourage students, widen their horizons and help them to transition successfully to the next stage of their life.

Throughout the book, the definition of career that I use is an individual's 'journey through life, learning and work' (Andrews and Hooley, 2022). This is taken from the *Careers Leader Handbook*, also published by Trotman, which is used by the careers teams in many schools in England.

There are thousands of STEM careers, and this book can't list them all, but the examples and stories included will help young people with their careers research.

How you can use this book

Each chapter includes an introduction, a conclusion and finishes with a reflection activity. The main content of each chapter is 'chunked' to make it easy for the casual reader to dip in and out. (See Chapter 1 for an overview of each chapter.) Please visit the **Trotman website** using the QR code or URL at the start of this book for discussion ideas and activities that you can use with your students, along with lesson resources that support the 'Heroes in STEM'.

Other features of the book

Heroes in STEM

Six fairly well-known scientists, technologists, engineers and mathematicians are also featured in the book. Each piece can be used to start a discussion about the sector/career/study route. **There is an accompanying lesson activity idea for each 'hero' on the Trotman website**. They are not prescriptive lesson plans but are intended as ideas for you to adapt to your students' needs. You can incorporate the activities into your department's career learning plan and perhaps add the students' work to a STEM careers display.

Fascinating facts

Ten 'fact boxes' are included in the book, demonstrating an innovative area of STEM and one or two associated careers. This can be a good starting point for research about the topic and its application or the career and what training and qualifications are required. STEM subjects are full of fascinating facts, and you never know what might ignite a young person's interest, so choose your own facts to share in the classroom. However, remember to link to a sector, job or skill to help start a career conversation.

Career stories

Interviews with young people from across the UK who are in their early STEM careers are included in the book. For contrast, there is a career story from a mathematican transitioning from full-time work to semi-retirement. There are a range of careers across many sectors, including the creative arts, with stories of their day-to-day work. However, the personal stories of gaining experience, balancing life, overcoming setbacks, changing plans and learning to adapt are arguably just as important for young people to read about. References to these 'career stories' are made throughout the book, illustrating why the advice in the book is useful and important. Please encourage your students to read all the interviews, as our contributors have some great lessons in life to share!

Creative industries

Although the book is called *STEM Careers*, the creative industries are included. In the world of work, disciplines are not as demarcated as they are in education. Technology and engineering, science and maths are needed in our vibrant gaming, fashion and media industries. Read about Molly, the podcast producer, in Chapter 2. Her stories about media production highlight the overlap between STEM and the creative industries. Thanks to the developers of the ERIC app for their guidance in including creative industries information in the book. Visit their website (www.meet-eric.com/creative-industries).

Website resources

Collections of quality-assured classroom resources

www.stem.org.uk/secondary/careers
https://resources.careersandenterprise.co.uk/ (search for STEM or your subject)

Other ways you can help your students

- Encourage students to engage with the school/college careers adviser. If you can, help each young person get to know the careers adviser by inviting this person into your classroom to give an overview of STEM study routes or local STEM employers. For those students not interested in STEM, ask the careers adviser to talk about why all employers value skills developed in STEM subjects, such as problem-solving, numeracy and digital skills.
- Develop cross-curricular opportunities with non-STEM colleagues for students to use skills and knowledge from both subjects, for example, working with geography when discussing renewable energy: What are the issues in relation to land use? How does that fit with the STEM understanding of the technologies? Another idea is to collaborate with the English department to develop persuasive writing skills about technological or environmental issues and possible solutions (see the online resources activity linked to the Hero in STEM, Hamza Yassin). Choose a current topic that you know your students are interested in and encourage career exploration from both STEM and non-STEM perspectives. Cross-curricular collaboration can also be a great opportunity to work with the visual arts departments, as the arts play an important role in widening public understanding of local and global issues. An example of an activity is to work with the art department to design and create a display to show the problems, and solutions, of waste plastics.
- Provide opportunities for your students to become their own careers researcher. Encourage them to plan a course of action, evaluate success and adjust goals, for example, improving an exam grade or developing practical skills. This will help them to exercise their agency, developing behaviours that will help them in adult life.
- Most importantly, help your students believe in themselves and develop their self-efficacy:
  - Give them opportunities to practise researching STEM careers and thinking about how to action plan for a specific job or career pathway.

- Encourage them to find career stories from people working in sectors they are interested in and help your students see they are developing the skills desired in the workplace.
- Allow time for students to discuss what they have learnt from careers-related experiences, for example, careers research, school trips, work experiences and any part-time work. Provide positive feedback that is task-oriented rather than ego-oriented for the learning young people gain.

Conclusion

This chapter is for all teachers. We can help students understand the interdisciplinary aspect of many careers. For example, to become a town planner may require geography, maths and IT skills, but an eye for detail, an aesthetic flair and a good understanding of human behaviour will also help. Therefore, other subject teachers may find it useful to dip into this book.

The teacher has an influential role to play in helping young people to see that a career in STEM can be for them and in recognising the value of the skills they have learnt in STEM subjects, regardless of their career aspirations.

Pathways are not always linear, and neither are our careers. Think back on your own career: you may have had part-time work while in education, taken part in volunteering activities, had a disrupted education and changed jobs and even sectors. All of this forms your 'career', contributing to your knowledge and skills, affecting your behaviours and decisions made.

Career – 'an individual's pathway through life, learning and work' (Andrews and Hooley, 2018).

Use this definition of career to guide the work you do with young people, helping them to begin to manage their career, seeing that learning is lifelong and that our life outside of work is important.

I hope you enjoy using this book and the accompanying lesson resources on the website.

A-Z 60 additional STEM career ideas

There are 60 sector-specific jobs discussed in Chapters 3 and 4. Here are 60 more jobs, in alphabetical order, chosen to represent more STEM occupations. These illustrate how broad the range of STEM occupations are while hinting at how specialised in one area it is possible to become.

Remember, this is just a list. Please use it to help you research further and perhaps find other career areas that may interest you. If you are unsure about your future career plans, read the initial advice in Chapters 1 and 2 and follow the reflection activities suggested at the end of each chapter.

Looking for more STEM career inspiration? You will find lots more STEM jobs and careers information at the following websites.

www.destinationstem.org.uk/
https://icould.com/
www.prospects.ac.uk/job-profiles
www.ucas.com/explore/career-list

STEM Careers

 **Accountant**

Accountants work for organisations or individuals to ensure their finances conform to regulations and provide them with a good rate of monetary return. They examine how money has been spent or invested and how it might be in the future. The profession operates in the public and private sectors.

 Acoustician

Acousticians use their knowledge of the physics of sound to better manage it. This involves improving sound quality, reducing noise levels and helping organisations to find the causes of unwanted noise and solutions to it.

 Actuary

Actuaries calculate the likelihood of future events to enable organisations to estimate the risk levels essential to setting insurance premiums. They need a strong mathematical aptitude to understand and apply theories of probability and make sophisticated use of statistics.

 Aeronautical engineer

Aeronautical engineers use scientific principles and a deep knowledge of aerodynamics and mechanical engineering to design, construct and maintain aircraft. The majority work on planes, but some work on radar, missiles, satellites or even space vehicles.

 Aerospace engineering technician

Aerospace engineering technicians are mostly involved in building and testing aircraft, and repairing and servicing them between flights. They use sophisticated equipment to detect wear and tear and may specialise in small planes, large passenger jets or helicopters.

 Agricultural engineer

Agricultural engineers design, develop, test and manage equipment used in farming, horticulture and forestry, and in maintaining recreational surfaces such as sports fields and golf courses. Much of the machinery they make is used to cultivate soil, harvest crops, store land produce or feed livestock.

A–Z of STEM careers

 Air quality engineer

Air quality engineers use chemical analysis and computer modelling to monitor air pollutants and the impact of industry and transport on the quality of the air within specific locations. This is an important role as poor air quality has been linked to many health issues.

 Analytical scientist

Analytical scientists normally work in laboratories. They work out the exact nature of chemical substances. Their roles can involve helping to protect the environment, ensuring high standards of food, drink and drugs; diagnosing disease; and guaranteeing the safety of manufacturing processes.

 Archaeologist

Archaeologists study the human past, initially through physical remains such as bones, pottery, tools and buildings. Excavating remains must be done with great care to preserve them in the best possible state for examination and analysis. Many archaeologists are employed by museum services, universities or large organisations devoted to preserving the past, such as English Heritage and Historic Environment Scotland.

 Architect

Architects plan and design buildings and are involved in their construction until completed. Their commission may be for a single structure (like a school or library), many similar buildings (such as a housing estate) or to renovate one, such as a listed building.

 Astronomer

Most astronomers work at a university or observatory studying the stars and planets, and other celestial phenomena. Typically, they use equipment such as telescopes and satellites to collect data (called 'observational' work) or employ computer models to test ideas of what happens in space (known as 'theoretical' work).

151

STEM Careers

 Biochemist

Biochemists study life processes at the molecular level in all manner of organisms, from visible ones (such as plants and animals) to those seen only under a microscope. They may work in industry, helping develop new, safe foods; in agriculture, improving fertilisers or pesticides; or in medicine, analysing body tissues and fluids to aid in diagnosing disease.

 Chemical engineer

Chemical engineers are knowledgeable about the processes which cause changes in the composition of substances. Their expertise is essential to converting raw materials into complex products such as fibres, plastics, paints, drugs, dyes and cleaning agents.

 Colour technologist

Colour technologists research and develop dyes and pigments (collectively known as colorants) or are involved in the actual process of colouring the materials themselves, which include paper, textiles, fibres, food products, detergents and soap.

 Computer game tester

Computer game testers play these games to ensure they work properly, often doing this for different levels or versions of the product. They record any problems or 'bugs' they find and may also check that what's said in instructions and on packaging is correct, compare the game with similar ones on the market and recommend possible improvements.

 Control engineer

Control engineers research, design and manage the equipment which guides and monitors machinery, usually in the form of electronic or computer technology. They need to understand the nature and limits of human capacity, as well as how machines operate.

 Dental technician

Working to the instructions of a dentist or doctor, dental technicians make and repair dental appliances, the majority being crowns, bridges and dentures. Great precision is important, and they use a wide range of materials, including metal alloys, plastics and ceramics.

A–Z of STEM careers

 Dentist

Dentists mostly see patients for regular check-ups to prevent tooth decay or gum disease. Treatments include scaling and polishing teeth, drilling and filling cavities, and extractions. Dentists also straighten irregular teeth and fit crowns, bridges or dentures. They are assisted by dental nurses and may also supervise dental hygienists or therapists. Most dentists work in high-street practices, but some are hospital-based, where they are usually employed to offer a specialism.

 Doctor

Doctors diagnose and treat ill health and (where possible) act to prevent its occurrence. Most are either GPs or hospital doctors, the latter including numerous specialisms. Most doctors need excellent communication skills because they need to offer information and advice to patients. Most also belong to a team, which may include nurses, radiographers, pharmacists or dieticians. Some doctors become researchers and may have little or no contact with patients.

 Drone technician

Drone-related jobs are still developing, but the use of drones is increasing in many sectors, including inspection/observation work and transportation. Being able to pilot, maintain and programme drones will become increasingly important roles. Drone pilot is a relatively new career choice that could offer employment in a range of sectors, from agriculture and construction to conservation and entertainment.

 E-learning designer

E-learning designers produce software for educational and training purposes, often for users without an instructor. The packages they create need to be self-explanatory, as well as interesting to use. Besides technical knowledge, they need to understand how people learn or work with a specialist who does.

 **Electrical engineer**

Electrical engineers research and design telecommunications systems, computers, satellite systems and television, while electrical engineers ensure the generation and supply of electricity for domestic, commercial and industrial purposes. Some act as consultants and may train their customers to use new systems properly.

STEM Careers

 Engineering draughts person

There are two stages in the engineering drawing process – 'design' and 'detail'. The first involves making a scale drawing showing the general outline, along with details of the nature and number of parts required, while the second consists of producing the very precise drawings which the production-line staff refer to at each stage of the manufacturing process.

 Food quality analyst

Food quality analysts (or food technicians) are usually laboratory-based, where they help develop food products and test them for quality and safety. As well as food features such as taste and colour, they check raw ingredients and storage methods to safeguard against dangerous micro-organisms like salmonella and *Escherichia coli*.

 Forensic scientist

Forensic scientists provide scientific evidence for use in courts of law. They may work within a police force but do so as civilian employees and must approach their tasks impartially. These tasks are mostly done in a laboratory environment but are sometimes at the scene of a crime. As well as providing evidence in a written report, they may be cross-examined in court.

 Gas network engineer

Usually working as part of a team, gas network engineers install gas supply to houses, industrial and commercial premises, schools and hospitals, and other public buildings. Having installed it, they ensure the safe continuance of the supply and provide an emergency service as needed. Many engineers are retraining so they can install hydrogen technologies.

 Geologist

Geologists study rocks, minerals, fossils, crystals and sediments to learn more about our planet and its resources. The potential applications of their knowledge include oil and gas exploration, mining, geological surveying and civil engineering.

A–Z of STEM careers

Geophysicist

Geophysicists study the physical make-up, motion and other workings of the earth. Their activities often result in very precise mapping of specific features or areas, which can be essential in locating and extracting mineral resources, or those vital to life itself, like water. Sometimes they can act as specialists within another occupation, like archaeology.

Horologist

Horologists repair and restore watches and clocks, most specialising in one or the other. This entails dismantling and cleaning the mechanism, and repairing or replacing damaged or worn parts. They use small tools, including eyeglasses and tweezers, to operate very precisely. Some watch repairers specialise in mechanical items, some in quartz and some in electronic.

Hydrologist

Hydrologists monitor and manage water resources in a range of settings and use them for different purposes, some domestic, others industrial, commercial or environmental. They ensure an effective flow of water through pipes or channels and contribute to the effective planning and sustainability of water resources. Some are involved in environmental work and river habitat restoration.

Laboratory technician

Laboratory technicians are employed in a wide range of areas, including industry, education, research, medicine and commerce. Most help professionals engaged in science-related work, and their regular duties normally include setting up equipment, preparing and carrying out experiments, taking measurements and writing reports or summaries, as well as disposing of waste and ordering fresh stocks.

 Map-maker

Map-makers (also known as cartographers) prepare maps and sea charts, globes of the earth's surface, three-dimensional models of geographical areas and even representations of the night sky. Besides existing maps, they also use aerial photographs or seismic sensing to produce finished graphics, whose nature will depend largely on whether they're intended for non-experts or professionals.

Marine biologist

Marine biologists study sea creatures and the environment on which they depend to raise our understanding of the marine world and to predict changes in ecosystems which may prove significant. They mostly work close to the coast or on the open sea and contribute to conservation, education and research. A project could take them anywhere in the world.

Maths teacher

Teachers can operate as maths specialists from the beginning of the secondary level (Year 7 or equivalent). Maths is a compulsory subject up to and including GCSE (or equivalent), so most of them teach to all levels of ability. The main areas taught are called number (arithmetic), algebra, geometry, probability and statistics.

Meteorologist

Meteorologists study the sciences of weather and climate to make short- and long-term forecasts. They do this by collecting and interpreting abundant data relating to atmospheric pressure, temperature, humidity, wind and clouds in order to build computer models geared to this. A few become well-known as weather presenters on TV.

Nanotechnologist

Nanotechnologists design and build devices on a tiny scale. They normally work in a laboratory, wearing protective clothing and using specialist scientific equipment. They're mainly employed in industries such as electronics, energy production and storage, automotive, aerospace, biotechnology, medicine and pharmaceuticals, and food science and production.

Nuclear engineer

Nuclear engineers design nuclear plants, supervise the manufacture of the equipment used there and operate the finished installations. They must guarantee a safe environment in the installation and surrounding area, and shut down the plant when required for maintenance or inspection, or in an emergency. They retrieve, treat and store radioactive waste from the site and develop waste management strategies, which have to be for the very long term.

A–Z of STEM careers

 Nurse

Nurses work in one of four specialist branches: adult nursing, mental health, learning disability and children's nursing. They look after people who are unwell or who need special care, sometimes in the long term. They do practical tasks like dressing wounds or administering drugs, but also plan how to meet their patients' needs. Many work in hospitals, but also in GP surgeries and community clinics. Within the four main branches, there are specialisms such as intensive care, cancer care and theatre nursing.

 Nutritionist

Nutritionists are knowledgeable in the science of how our bodies use food and the way this relates to good health and illness. They advise people on eating healthily in order to avoid serious conditions such as heart disease, cancer, stroke or diabetes. Many work in the NHS or the food industry (in research or labelling), but they can also be employed in the community, for example, coordinating healthier eating programmes. Some work within sports, as personal trainers or part of a competitive professional team.

 Oceanographer

Oceanographers study the earth's oceans and seas and how these interact with the atmosphere and land masses. Most of their data are collected on the water, using research vessels, instruments or floats, or robotic devices. They use this knowledge to promote the responsible use of marine resources and to minimise environmental damage.

 Offshore windfarm site manager

Offshore windfarm site managers are responsible for managing the site, which includes the safety of the personnel and the maintenance of the wind turbines. They need good technical knowledge to understand the windfarm infrastructure and excellent people skills for human resources management.

STEM Careers

 **Orthotist**

Orthotists design and fit surgical devices, such as spine supports and braces, to support part of a patient's body, usually to relieve pain or act for paralysed muscles. They also make any adjustments, repairs or renewals as the need arises. They need a good understanding of the musculoskeletal systems and of the materials and technology used to make the devices.

 Paleontologist

Paleontologists study fossils to build their knowledge of life forms which existed in the very distant past. The clues these provide about this past offer important data concerning environmental and climate change.

 Physiotherapist

Physiotherapists assess and treat people with restricted movement, typically caused by age, illness or injury. Most of the problems they address relate to joints or bones, but some pertain to the cardiovascular (heart and lung) and nervous systems. They work in hospitals, outpatient clinics, workplaces and older people's homes. GPs and other healthcare professionals often refer patients to physiotherapists.

 Plumber

Plumbers install, maintain and repair hot and cold water systems, central heating and drainage, and the pipes and controls used for gas supply. They may work in houses and commercial or industrial premises, often needing to consult complex plans or drawings before taking action. They need considerable skill to properly use a wide variety of tools to cut, bend and join metals and plastics.

 Public health intelligence

A public health intelligence professional is responsible for gathering data relating to public health, interpreting this information and planning strategies to deal with issues that affect people's health. Many work for the National Health Service, and there is also the opportunity to work with global operations, such as the World Health Organization.

A–Z of STEM careers

 ## Quantity surveyor

Quantity surveyors calculate the cost of building projects, taking account of materials, labour and maintenance. Many work for commercial contractors and are based on a site, usually for housing. However, some are employed by local authorities, where they're more likely to engage in single-structure programmes, whether new or for renovation.

 ## Radiographer

Radiographers work in either diagnostics or therapeutics. The former produce precise images of the body which aid in the diagnosis of conditions and monitor the progress of a condition or a patient's rate of recovery. They use X-rays, magnetic resonance imaging (MRI), ultrasound and computed tomography to build two- and three-dimensional images. Therapeutic radiographers work as part of a team treating cancer patients, operating with great precision to deliver exact doses of radiation to affected body areas.

 ## Science teacher

Most specialist science teachers work in secondary schools, covering one or more of biology, chemistry, physics and geology. Like English and maths, science is compulsory to GCSE, and many students take one or more from this group to A level (or equivalent). Skill is needed in using science equipment, and some projects may require fieldwork. Health and safety considerations are always a priority.

 ## Speech and language therapist

Speech and language therapists work mainly with children and adults who have problems producing or understanding speech, suffer from a stammer or have difficulty swallowing. These issues often result from hearing loss, stroke, disability, injury or a degenerative condition. Therapists must quickly establish a trusting relationship with each client to make an accurate diagnosis leading to effective treatment.

STEM Careers

 ### Statistician

Statisticians collect, analyse and interpret numerical data. They must act neutrally, presenting actual findings arrived at by respected methods, never being swayed by what they (or those who employ them) might prefer the results to be. Many work for the civil service or in local government.

 ### Sustainability manager

A sustainability manager oversees the implementation of sustainability strategies for an organisation or business. This may involve analysing the processes of the company and planning strategies to reduce energy consumption and increase efficiency. They will have a background in environmental work and business.

 ### Surveyor

While some surveyors are employed in specialist fields, most work in land or property, taking measurements and making estimates of value. Many are involved in house sales, where they check properties for structural damage or instability and draw attention to any hazards or shortcomings in functions.

 ### Telecommunications engineer

Telecommunications engineers try to improve existing communications technology and to develop new products. The focus of their work is on areas like internet connectivity, mobile phones, cable and satellite systems, and coordinating telephone and computer systems.

 ### Theatre sound technician

Theatre sound technicians set up and operate sound equipment for live productions such as plays and musicals. They must maintain the equipment, which in a touring production may need to be dismantled after each show. Setting this up to produce good-quality sound requires considerable expertise, as every theatre is different.

A–Z of STEM careers

 Urban planner

Urban planners may have a geography or construction background. They help to develop projects linked to sustainable housing and road building, and the development of urban spaces to benefit local communities. They need good numerical skills and an understanding of human behaviour.

 Vehicle technician

Vehicle technicians service, overhaul and repair light vehicles (such as cars and motorcycles) or heavy ones (such as buses or trucks). They're very skilled at diagnosing faults and often work under pressure to quickly repair vehicles needed promptly. Some technicians specialise in construction plants, such as diggers or cranes.

 Virologist

Virologists study viral infections like hepatitis and HIV. Many work within a clinical microbiology service, identifying the nature of viruses which cause infection. They may screen selected populations at risk from specific viral diseases and sometimes investigate the spread of infections by scrutinising the design and maintenance of clinical areas (like operating theatres), food preparation and hygiene or cleaning and waste disposal procedures.

 Welder

Welders use intense heat to join together pieces of metal, often operating in a workshop or on a construction site. Normally, they use an electric arc or a gas flame and wear protective clothing, including a helmet, thick gloves and boots, and ear protectors. In some industries, like car manufacturing, most of the operations are done by robots, but welders set up the machines and oversee the work.

 Zoological scientist

Zoological scientists study animals in the wild and in captivity to help them live well and ensure their long-term survival. Many are attached to zoos, safari parks or universities, though project work often takes them abroad, sometimes to remote and exotic places. Technological developments have now made observation of animal behaviour much easier and less obtrusive.

NEW AND BEST-SELLING FROM TROTMAN

New in 2024

 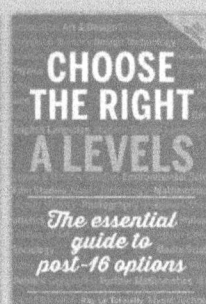  

Get into University

 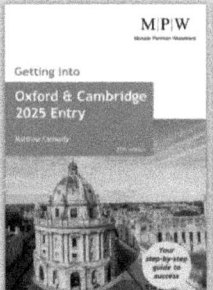  

Careers Essentials

   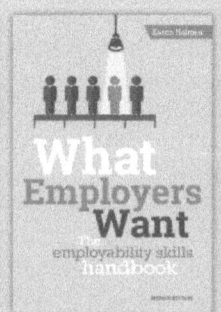

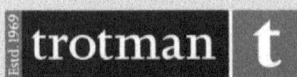

Enhance your careers library with our bestsellers, visit:
www.trotman.co.uk

www.ingramcontent.com/pod-product-compliance
Lightning Source LLC
Chambersburg PA
CBHW061252230426
43664CB00025B/2934